# LABORATORY MANUAL

# ANATOMY & PHYSIOLOGY

## TENTH EDITION

**KEVIN T. PATTON, PhD**
*Professor of Anatomy & Physiology Instruction*
*New York Chiropractic College*
*Seneca Falls, New York*

*Founding Professor of Life Sciences, Emeritus Faculty*
*St. Charles Community College*
*Cottleville, Missouri*

*Assistant Professor Emeritus of Physiology*
*Course Director Emeritus in Human Physiology*
*St. Louis University Medical School*
*St. Louis, Missouri*

### LEAD CONTRIBUTOR
**FRANK B. BELL, DC, MSHAPI**
*Associate Professor of Biology*
*SUNY Adirondack*
*Queensbury, New York*

*Assistant Professor of Human Anatomy and Physiology Instruction*
*New York Chiropractic College*
*Seneca Falls, New York*

*Adjunct Instructor*
*Excelsior College*
*Albany, New York*

### CONTRIBUTORS
**DANIEL J. MATUSIAK, EdD**
*Adjunct Professor*
*St. Charles Community College*
*Cottleville, Missouri*

*Life Science Teacher*
*St. Dominic High School*
*O'Fallon, Missouri*

**STEVEN R. WOOD, MS, MA**
*Professor of Anatomy and Physiology (adjunct)*
*Lone Star Community College—University Park*
*Houston, Texas*

*Professor of Biological Sciences (retired)*
*Florida State College at Jacksonville*
*Open Campus—Deerwood Center*
*Jacksonville, Florida*

ELSEVIER

# ELSEVIER

3251 Riverport Lane
St. Louis, Missouri 63043

ANATOMY & PHYSIOLOGY LABORATORY MANUAL
TENTH EDITION

ISBN: 978-0-323-52892-4

---

**Notices**

Practitioners and researchers must always rely on their own experience and knowledge in evaluating and using any information, methods, compounds or experiments described herein. Because of rapid advances in the medical sciences, in particular, independent verification of diagnoses and drug dosages should be made. To the fullest extent of the law, no responsibility is assumed by Elsevier, authors, editors or contributors for any injury and/or damage to persons or property as a matter of products liability, negligence or otherwise, or from any use or operation of any methods, products, instructions, or ideas contained in the material herein.

---

Previous editions copyrighted 2016, 2013, 2010, 2007, 2003, 1999, 1996, 1993, and 1987.

**International Standard Book Number: 978-0-323-52892-4**

*Executive Content Strategist:* Kellie White
*Director, Content Development:* Laurie Gower
*Senior Content Development Specialist:* Laura Goodrich
*Publishing Services Manager:* Julie Eddy
*Book Production Specialist:* Clay S. Broeker
*Design Direction:* Brian Salisbury

Printed in Canada

Last digit is the print number:  9  8  7  6  5  4  3

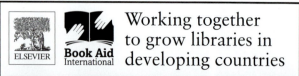

# Contents

# Preface

Anatomy and physiology laboratory courses provide the essential hands-on learning opportunities required for a thorough understanding of the human body. This manual contains a series of 55 exercises that provide several guided explorations of human structure and function to accompany any standard human anatomy and physiology textbook. These activities include:

☐ **Labeling exercises** provide opportunities to identify important structures learned in the laboratory and lecture portions of the course. Once completed, the exercises provide guidance for laboratory examinations of models and specimens. Students are encouraged to write out the labels so that the terms are more easily learned.

☐ **Coloring exercises** are becoming the most popular and effective way for many learners to grasp the essential spatial relationships of anatomical structures. This manual contains an accurate and comprehensive collection of anatomy and physiology coloring plates.

☐ **Dissection of anatomical models** and examination of charts are an integral part of any beginning anatomy and physiology laboratory experience. This manual's instructions give valuable guidance for the effective use of models and charts.

☐ **Dissection of fresh and preserved specimens** of tissues, organs, and whole organisms enhances student appreciation of anatomical and functional relationships. Examination of tissues and organs such as those from sheep and cows is suggested in specific exercises throughout this manual. The rat, the cat, and the fetal pig are offered as options for many of the more comprehensive dissection activities. The last unit offers a guide to a live or taped demonstration with a prosected human cadaver.

☐ **Physiological experiments** emphasizing a variety of functional processes of the human body offer students immediate and dramatic examples of physiological concepts. When possible, these activities center around examination of the student's own physiological processes.

☐ **Content and concept review** questions and fill-in tables throughout the manual encourage students to reinforce and apply their knowledge of human structure and function.

☐ **Modern anatomical imaging** techniques such as computed tomography (CT), magnetic resonance imaging (MRI), and ultrasonography are introduced where appropriate. In each special presentation of imaging technology, students are challenged to interpret actual images of the human body.

☐ **Practical applications** for exercise and athletics, clinical situations, and everyday experiences increase student motivation and place important concepts in a useful context. Each practical example includes application questions that encourage students to think about how concepts apply to the situation described.

This laboratory manual offers other features that enhance learning and ensure a safe and effective laboratory experience:

☐ **A full-color design** enhances the learning experience by making the material in the lab manual easier to navigate and use.

☐ **Learning objectives** presented at the beginning of each exercise offer a framework for learning.

☐ **Complete lists of materials** for each exercise give the students and instructor a handy reference for efficient setup of laboratory activities.

☐ **Boxed hints** provide students with special tips on handling specimens, using equipment, and otherwise managing their laboratory activities.

☐ **Safety tips** are highlighted in special boxes to remind students of potential hazards such as fire, chemical spills, cuts, and biological contamination.

☐ **Study tips** are highlighted in special sections to assist students in their study of specific topics in this manual.

☐ **Numerous illustrations** of proper procedures complement the text's complete description of laboratory activities. All anatomical illustrations include a handy anatomical compass rosette to orient students to anatomical directions and to reinforce development of spatial perspective.

The design of this laboratory package not only makes the laboratory course fun and effective for the student, but it also provides essential support for the instructor and lab preparation technician. Here are some examples of elements designed to aid instruction and preparation:

☐ **A comprehensive instruction and preparation guide** is provided to each adopting instructor. The guide contains a complete set of hints, special notes, and instructions for each laboratory exercise. The guide provides a list of materials broken down by exercise and a comprehensive list of all materials suggested for the course. Substitutions and special sources are given where appropriate. Lists of solution-preparation guidelines and other aids are found throughout the guide. Reproducible handouts to supplement certain exercises are also provided in the guide.

☐ **Modular organization of laboratory exercises** allows their use in virtually any order required by the needs of individual courses. Comprehensive cross-references in the instruction and preparation guide alert instructors about other lab exercises that may involve similar material. For example, Lab Exercise 33: Hormones may be more appropriately done near the reproductive system exercises in some courses.

☐ **Complete instructions for lab activities** are given to the student, freeing the instructor to interact with laboratory students on an individual basis rather than spending a great deal of time introducing the lab procedure to the whole class.

The tables in this section give the basic units and common alternate units for time, temperature, length, volume, mass, and pressure.

The abbreviations of metric units are given in parentheses. Note that they do not have a period (.) after them. Be careful to notice whether they are capital (uppercase) letters.

### TIME

Basic unit: second (sec)

0.000001 second = microsecond (µsec)

0.001 second = millisecond (msec)

60 seconds = minute (min)

3600 seconds = hour (hr)

### TEMPERATURE

Basic unit: degree Celsius (° C)

No alternate units are commonly used.

### LENGTH

Basic unit: meter (m)

0.000000001 meter = nanometer (nm)

0.000001 meter = micrometer (µm)

0.001 meter = millimeter (mm)

0.01 meter = centimeter (cm)

1000 meters = kilometer (km)

### VOLUME

Basic unit: liter (l or L)

0.001 liter = milliliter (ml)*

0.01 liter = centiliter (cl)

0.1 liter = deciliter (dl)

*milliliters = cubic centimeters (cc)

### MASS

Basic unit: gram (g)

0.001 gram = milligram (mg)

0.01 gram = centigram (cg)

1000 grams = kilogram (kg)

### PRESSURE

Basic unit: millimeters of mercury (mm Hg)

No alternate units are commonly used.

Accurate measurement means using units of measurement correctly, but it also means using measuring devices accurately. If you are not familiar with reading the markings on metric rulers, balances, thermometers, and other common measuring devices, ask your instructor to demonstrate.

## HOW TO DISSECT

The term anatomy literally means "to cut apart," so it is no wonder that dissections are commonly done in anatomy lab courses. Dissection activities recommended in this manual are not proposed without recognition of humane concerns. The purpose of these dissections is to instruct in a way that no other method can duplicate. For health professionals, dissection of anatomical preparations is a necessary prerequisite to working with living bodies.

The goal of any dissection exercise is the exploration of anatomical relationships. Proper dissection requires patience and skill. Some students slice and hack away at their specimens; others hardly touch them. Good technique lies somewhere between these two extremes. Organs should be separated from one another only enough to see surrounding structures. Rarely should structures be cut or removed. The instructions given in this manual state when, where, and how cuts should be made.

Each dissection activity in this manual offers safety advice concerning proper handling of the dissection specimens. Protective gloves, lab coat, and eyewear are recommended. Always be careful when using dissection tools; severe injuries can result from their careless use.

**Figure A** shows some commonly used dissection instruments. A brief discussion of each follows:

☐ 1 **Scalpel or knife.** The scalpel is probably the instrument most overused by beginners in an anatomy lab course. This tool should rarely be used and only when you want to cut all the way through a specimen. A pathology knife, or butcher knife, is a much more useful tool.

☐ 2 **Scissors.** Scissors are probably the most underused dissection tool. Whenever you are tempted to use a scalpel for cutting, try the scissors first. Often, your results will be much better and you will not have damaged important underlying parts.

☐ 3 **Probes.** You may want to have several types of probes. Both dull and sharp probes (dissecting needles) are useful in separating tissues, exploring cavities, tracing blood vessels, and pointing to structures.

☐ 4 **Forceps.** Forceps are one of the handiest dissection tools. They can be used to grasp small objects, to separate structures, to point to structures, to explore cavities, and to pull on structures.

☐ 5 **Ruler.** Your metric ruler should be marked in both centimeters and millimeters. It is useful in measuring organs and in many non-dissection lab activities.

☐ 6 **Dissection pins.** These are standard, heavy-duty straight pins. Dissection pins are useful in pinning membranes and other structures to a dissection board to keep them temporarily out of the way.

☐ 7 **Dissection tray.** There are many varieties and shapes of dissection trays, and your instructor will recommend the best for your situation. Dissection trays help organize the dissection activity by keeping everything together, and they protect the lab table surface. Some types can also be used for storing your specimen for later study.

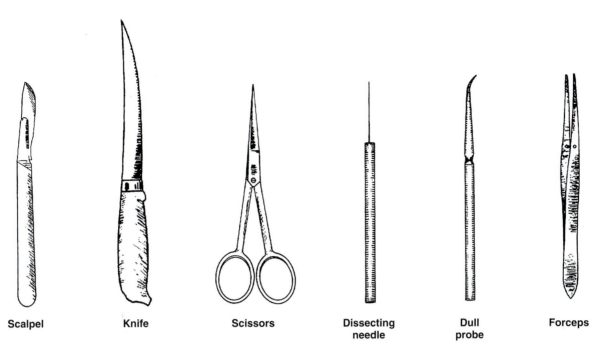

**FIGURE A**  Commonly used dissection tools.

## USING THIS MANUAL

The format of each Lab Exercise is self-explanatory. After some brief introductory remarks, each exercise begins with a section entitled "Before You Begin." This section recommends that before starting any lab activity you should:

☐ Read the appropriate chapter(s) in the textbook you are using for this course.

☐ Set your learning goals so that you know what important concepts you should be trying to learn.

☐ Prepare your materials so that your lab activity will run smoothly.

☐ Read the entire activity before starting. In this way, you will not be surprised (and therefore unprepared) for any step in the procedure.

☐ Each exercise contains one or more activities. Your instructor may suggest that you do all the activities in an exercise or only one. Each activity includes some or all of these helpful features:

☐ Large, bold step numbers help you keep your place as your eyes move back and forth between the manual and your lab setup. Each step number is preceded by a check-box (☐) that you can use to check off each step as it is completed.

    ☑ 1 Find these structures on your specimen:
       ☑ head
       ☑ neck
       ☑ torso
    ☑ 2 Carefully open the ventral body cavity of your specimen.
    ☑ 3 Locate the following in the ventral body cavity of your specimen:
       ☑ stomach
       ☑ spleen
       ☑ liver

☐ The following features highlight useful information for completing the activities safely and successfully:

 **HINT boxes**—Offer valuable information and dissection techniques.

 **SAFETY FIRST boxes**—Draw attention to important safety concerns pertaining to the lab activities.

**LANDMARK CHARACTERISTICS boxes**—Provide helpful advice for identifying specimens.

**STUDY TIPS sections**—Just as the name implies, provide ideas to assist you in your study of specific topics in this lab manual.

☐ Anatomical dissection illustrations and photographs are identified with a colored tab at the top of the page for quick and easy reference.

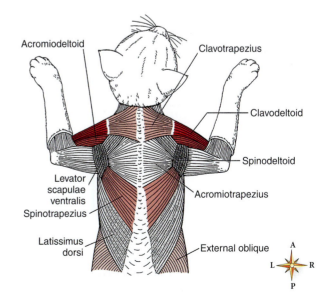

☐ Labeling exercises encourage you to apply your knowledge of human anatomy in a practical test. Long label lines allow you room to write the name of the structure directly on the illustration. When each figure is completely labeled, the terms should be transferred to the Lab Report so that the instructor can check your work.

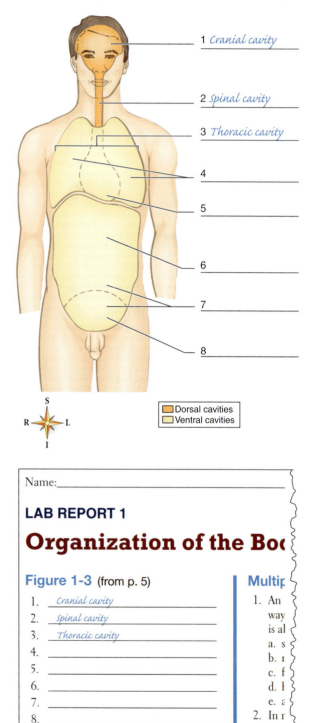

1 *Cranial cavity*

2 *Spinal cavity*

3 *Thoracic cavity*

4

5

6

7

8

☐ Dorsal cavities
☐ Ventral cavities

Name:_____

**LAB REPORT 1**

# Organization of the Bod

**Figure 1-3** (from p. 5)

1. _Cranial cavity_
2. _Spinal cavity_
3. _Thoracic cavity_
4. _____
5. _____
6. _____
7. _____
8. _____
9. _____

**Multip**

1. An
   way
   is al
   a. s
   b. r
   c. f
   d. l
   e. a
2. In r
   ner
   the

☐ The coloring exercise has been cited by educators and students as being an effective method for learning anatomy. By using multiple senses to trace and highlight shapes and relationships, you reinforce your knowledge of the human form. Each coloring

plate is like a paint-by-number activity. Each label has a small red number that corresponds to a red number on the illustration itself. Use a colored pen or pencil (not a crayon) to write the term over the outline letters of the label. This will reinforce your familiarity with the term and its correct spelling and makes the label a "color code" so you can later find the structure by its color. After filling in the label, use the same color to shade in the matching structure in the illustration. Use contrasting colors so that parts next to one another can be easily distinguished. Do not attempt to color everything in realistic colors; otherwise many parts will look the same. If you use pens, make sure they won't bleed through the paper. If you use pencils, you may want to insert a piece of plain paper into your manual over each completed coloring plate so that the colors do not rub off on the facing page. The materials list in the first exercise reminds you to bring colored pens or pencils to your work area, but it is assumed that you will have these for all the lab exercises.

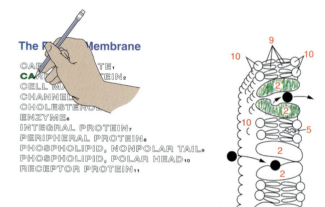

The            Membrane

CAR           TE₁
CAR           EIN₂
CELL M
CHANNEL
CHOLESTEROL
ENZYME₆
INTEGRAL PROTEIN₇
PERIPHERAL PROTEIN₈
PHOSPHOLIPID, NONPOLAR TAIL₉
PHOSPHOLIPID, POLAR HEAD₁₀
RECEPTOR PROTEIN₁₁

☐ Each exercise ends with a Lab Report. The report includes handy tables and sketching areas for you to record your laboratory observations and results. Where appropriate, the Lab Report also presents objective and subjective questions for you to answer. Some of these questions drill you on your knowledge of important terms, whereas others ask you to use your basic knowledge to interpret, organize, or otherwise process basic facts to ensure that you understand conceptual relationships. A number of highlighted practical application boxes in the exercise itself often ask you to apply your basic knowledge in this way. These questions, along with practice in making scientific observations, help you develop your scientific reasoning skills.

## SAFETY FIRST!

As already mentioned, numerous boxed safety tips appear throughout this manual. Entitled SAFETY FIRST, their large letters and urgent tone are meant to call your attention to potential hazards in the laboratory. Before beginning the course, it is essential that you learn some basic laboratory safety policies:

☐ Always use care and common sense in the laboratory.
☐ Check the labels of all chemicals for safety warnings before using them. If the container does not have a safety label, consult the instructor.

**Chemical Safety**

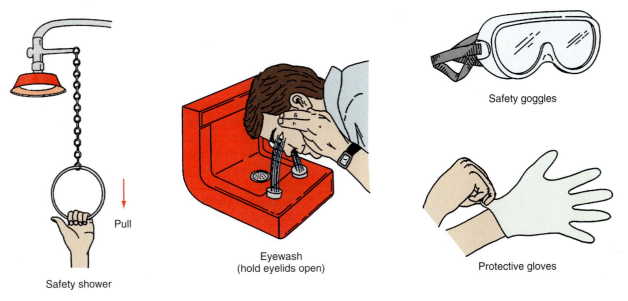

Pull

Safety shower

Eyewash
(hold eyelids open)

Safety goggles

Protective gloves

**Fire Safety**

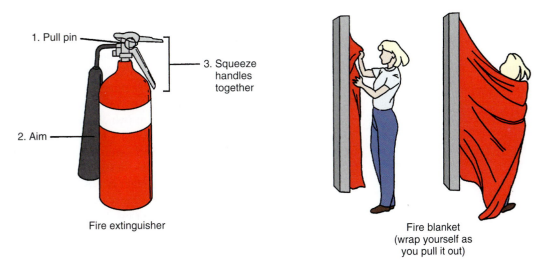

1. Pull pin

3. Squeeze
   handles
   together

2. Aim

Fire extinguisher

Fire blanket
(wrap yourself as
you pull it out)

**FIGURE B** Commonly used chemical and fire safety equipment.

☐ Wear protective gear if you are handling dangerous materials.
☐ Plan ahead about what to do if a spill occurs. If you spill a dangerous substance on your body, you may have to remove your clothing. If your lab has a safety shower, stand under it and pull the ring **(Figure B)**.
☐ Avoid using materials to which you have a known sensitivity.
☐ Locate first aid equipment in the laboratory and familiarize yourself with its use.
☐ Confirm all directions, procedures, and other information with your instructor before each activity. Mark changes or additions in this manual, crossing out any printed information that does not apply to you.
☐ Locate the nearest medical help and identify the easiest and fastest way to access it (for example, phoning an ambulance).

☐ Locate the fire safety equipment in your lab and familiarize yourself with its use. Locate the nearest fire alarm box and identify the recommended primary and alternative fire exit routes.

## Important Safety Information

(Use the spaces that follow to write important safety information.)

Location of first aid box:

Emergency medical help:

Fire extinguisher location:

Fire extinguisher type:

Fire alarm box location:

Primary escape route:

Alternate escape route:

Weather safety shelter:

Important notes:

☐ Always follow the directions and safety advice provided with equipment and supplies, even if they are different than those given in this manual. Injuries often result from the misuse of equipment.

☐ Remember that absolutely no eating, drinking, or smoking is allowed in the laboratory.

☐ To avoid injury or contamination, properly dispose of or clean and store all lab equipment and supplies before vacating the laboratory. Wash your hands thoroughly on entering and leaving a biological laboratory.

☐ Always supervise experiments or demonstrations in progress. Never leave a laboratory experiment unattended.

## HOW TO STUDY FOR LAB QUIZZES

Tips for effective studying could fill several volumes. Instead of an entire work on studying, this section presents a few tips that many students have found to be particularly helpful in the anatomy and physiology laboratory course. As with any list of study tips, you will find some more useful to you than others. You will find additional tips and practical advice in the *Survival Guide for Anatomy & Physiology* (second edition), published by Elsevier, and at *theAPstudent.org*.

☐ Identify the type of quizzes you will encounter. Written quizzes may involve multiple choice, matching, or other objective questions. Written quizzes may also have subjective short answer or essay questions in which you must interpret or restate concepts. Figures may be offered for labeling. Practical tests may involve stations at which specimens have been placed. You may be asked to perform a procedure or answer a question about the specimen while at each station. You may have an individualized practical quiz in which the instructor will ask you questions or watch you perform a procedure individually. Many lab courses involve online quizzes, lab reports, or other graded activities. Ask your instructor for some sample questions if you do not understand how an upcoming quiz is constructed. Knowing the manner in which you will be evaluated is important to designing your study strategy.

☐ Organize a study group (which may be the same as your lab group) to meet several times before a quiz to help each other with the material. Such groups can also operate by phone, email, discussion forums, or social media such as chat or group video. Quizzing each other is a useful technique.

☐ Spend many brief sessions reviewing specimens and other lab work, rather than one or two long sessions just before the quiz. Ask your instructor if there are open lab times or a learning center at which you can review models or specimens. Try to find online resources to help you practice.

☐ Study actively. Do not just sit and look at your study materials; do some hands-on activities with them. Make flash cards on paper or online to quiz yourself. Set up your own practical exam, and then take it. Sketch specimens and models again. Draw a "concept map" in which you draw boxes or pictures connected in a logical way. This will help you organize concepts in your mind.

☐ Use all of your resources. Your account at *evolve.elsevier.com* has learning resources. Campus libraries and learning centers often offer the services of specialists who can help you sharpen your study skills. They can show you examples of concept maps and flash cards. Your instructor has successfully completed laboratory courses and has helped numerous previous students in this course, so ask your instructor for study tips. Use the course textbook and the study aids in the textbook to help you learn the lab material.

# *the student laboratory*

The laboratory portion of the typical anatomy and physiology course is in some ways a unique beast. Here are a few ideas on how to make the student laboratory the best possible experience:

☐ **Safety is always first!** Scientists are sticklers for safety. They never take a step into the lab without every possible safety precaution being in place. A mistake in the lab, even a student lab, can cause serious injury or even death. Far too many preventable injuries occur in student labs. So PLEASE heed all of the safety precautions printed in your lab manuals, posted in your labs, and articulated or published by your school or professor or both. In addition, use common sense.

☐ **Use the "field method" for identifying anatomical structures.** Roger Tory Peterson, the famous naturalist and birder, developed a method for field identification of birds that has since been used for everything from identifying sparrows to spotting enemy aircraft. The Peterson method focuses on "field marks," the one or two characteristics that distinguish particular birds or airplanes from similar birds or airplanes. Simply approach tissues, bones, bone markings, muscles, and other structures that you need to identify in lab courses as you would animals in the field. Focus on the one or two characteristics that make one tissue different from all others. Peterson's method also makes use of range maps that show where different birds are likely to be found. Do the same with lab specimens. Where are adductor muscles likely to be found? On the inside of limbs. So, where would you start looking for adductor longus? Use a little tip from field geography too. It might take you all day to find the Mississippi River on a map if you have no clue what a river is. When learning bone markings, how can you find the supraorbital foramen when you have no idea what a foramen is? Study types of things, like types of muscles, types of bones, and so on, before studying the specific examples assigned to you.

☐ **Remember that the laboratory is a playground.** That doesn't mean you should just goof around for no good reason. It does mean that labs are intended to be fun, hands-on opportunities to play with specimens, models, and experimental equipment. Therefore, approach it that way. Don't just try to get by with the minimum activity. If you linger and look a little deeper, or turn the model around this way, then another way, you'll find a deeper understanding of what you were supposed to find as well as the connections that help you draw concepts together into the *big picture*. Kinesthetic (tactile) learners will especially benefit from this aspect of the lab course, often in ways that bring home points from the lecture or discussion portions of the course.

☐ **Use flash cards.** Flash cards are especially useful in certain lab activities. Many students like to draw pictures of lab specimens on one side of a card and its name or description on the other side. You might even bring a camera to take photos of your dissections so that you can use them to make flash cards! Your cell phone may have a camera, so use it as a learning tool in lab!

☐ **Use mnemonic devices.** Archeologists have recently uncovered evidence that mnemonic devices were used by the ancient Greeks thousands of years ago in their famous schools. You may have seen one of those memory experts that can spout off all the names on a page in the phone book using mnemonic devices. Mnemonic devices are little sayings or mental images you use to help you remember a concept. For example, the acronym *IPMAT* has been used by generations of students to help them learn the phases of the cell cycle: interphase, prophase, metaphase, anaphase, and telophase. The mnemonic sentence "On old Olympus's tiny tops, a friendly Viking grew vines and hops," is a mnemonic for learning the names, in order, of the 12 pairs of cranial nerves. (You can find these names in Table 24-1.) The sillier the phrase or mental image you use, or the more ribald, the easier it is to remember and the more useful it is as a memory aid.

☐ **Dissect like an artist.** Some of the best anatomists have in fact also been great artists. Artistic finesse in pulling apart structures without destroying them or removing them from their relative positions in the body is far more effective than simply hacking away. Your goal should be to find structures so that you can understand how the body is constructed.

You don't have to be a great sketch artist. However, you should *sketch a lot*. Draw everything! Don't sweat and slave over creating a masterpiece, even if you're capable. (Wait until after lab and work on your masterpiece on the weekend.) However, do take enough time that you can recognize key features later on when you use your sketches to study. In addition, label as much as you can. A sketch without any labels is pretty useless, except maybe as a coaster.

# *the student laboratory*—cont'd

☐ **Come to the lab prepared.** Read what you will be doing in lab before you do it. If you don't come prepared, then half the time in the lab will be spent just trying to figure out what to do. The other half of your time will be spent trying to rush through the activity without any appreciation of the concepts behind it. Preparation also means reading the portions of the textbook that relate to the lab activity. Preparation is probably the greatest key to using the student lab to its greatest potential.

☐ **Use many resources.** When you're trying to find anatomical specimens or interpret results from a physiology demonstration or experiment, don't rely solely on your textbook and lab manual. Often your specimen will not look like the picture in these books. No organ or tissue looks exactly like any other. Therefore, find out where you can purchase or borrow a lab atlas, anatomy atlas, histology book, and other useful resources. In addition, many useful resources for the anatomy and physiology lab can be found on the Internet. (You can use a search engine to find them.)

# PHOTO AND ILLUSTRATION CREDITS

**3-1:** Courtesy VEE GEE Scientific.

**6-3:** Courtesy Dennis Strete. **6-4:** Modified from Young B, O'Dowd G, Woodford P: *Wheater's functional histology*, ed 6, Philadelphia, 2014, Churchill Livingstone.

**7-4, 7-8:** Courtesy Ed Reschke. **7-5:** From Tallitsch R, Guastafelli R: *Histology: an identification manual*, St. Louis, 2009, Mosby. **7-6, 7-9:** Courtesy Dennis Strete. **7-7, 7-10:** From Young B, O'Dowd G, Woodford P: *Wheater's functional histology*, ed 6, Philadelphia, 2014, Churchill Livingstone.

**8-4, 8-5, 8-6, 8-7, 8-8, 8-9, 8-10, 8-11, 8-12:** Courtesy Dennis Strete.

**9-1, 9-2:** Courtesy Ed Reschke. **9-3:** Courtesy Dennis Strete. **9-4:** From Gartner LP, Hiatt JL: *Color textbook of histology*, ed 3, Philadelphia, 2007, Saunders.

**10-1, A:** From Erlandsen SL, Magney JM: *Color atlas of histology*, St. Louis, 1992, Mosby.

**11-3:** From Gartner LP, Hiatt JL: *Color textbook of histology*, ed 3, Philadelphia, 2007, Saunders.

**12-1, 12-2, 12-3, 12-7:** From Abrahams P, Marks S, Hutchings R: *McMinn's color atlas of the head and neck anatomy*, ed 4, Philadelphia, 2010, Mosby.

**13-1, 13-2:** From Muscolino JE: *Kinesiology: the skeletal system and muscle function*, ed 2, St. Louis, 2011, Mosby.

**14-1:** Courtesy Vidic B, Suarez FR: *Photographic atlas of the human body*, St. Louis, 1984, Mosby. **14-2, A, B:** From Muscolino JE: *Kinesiology: the skeletal system and muscle function*, ed 2, St. Louis, 2011, Mosby.

**15-1, A, B:** From Muscolino JE: *Kinesiology: the skeletal system and muscle function*, ed 2, St. Louis, 2011, Mosby. **15-2:** Courtesy Vidic B, Suarez FR: *Photographic atlas of the human body*, St. Louis, 1984, Mosby.

**17-2:** From Muscolino JE: *Kinesiology: the skeletal system and muscle function*, ed 2, St. Louis, 2011, Mosby.

**27-3:** Courtesy Ed Reschke.

**32-3, A:** From Kierszenbaum A: *Histology and cell biology*, Philadelphia, 2002, Mosby. **32-3, B, C:** Courtesy Ed Reschke. **32-3, D:** From Erlandsen SL, Magney JM: *Color atlas of histology*, St. Louis, 1992, Mosby.

**34-6:** Courtesy Biomedical Polymers.

**39-1, B:** Courtesy Photo Science Library. **39-1 C, D, E:** Courtesy Ed Reschke.

**49-2:** From Brunzel NA: *Fundamentals of urine and body fluid analysis*, ed 2, Philadelphia, 2004, Saunders.

**50-3, A:** From Brunzel NA: *Fundamentals of urine and body fluid analysis*, ed 2, Philadelphia, 2004, Saunders.

**55-5:** Courtesy Vidic B, Suarez FR: *Photographic atlas of the human body*, St. Louis, 1984, Mosby.

# Organization of the Body

Many explorers use maps, figures, and photographs to help orient themselves to the terrain to be explored. The anatomist uses maps, figures, and photos to explore the body and its parts. The use of maps and other aids to find a geographical position is called *orienteering* and is a useful analogy to human anatomical study. At the beginning of this exercise, you will learn how anatomical "maps" and models are read. Later in this exercise, you will become familiar with the major body systems and some of their organs so that you will be comfortable with the "lay of the land" in the human body.

## BEFORE YOU BEGIN

☐ Read the appropriate chapter in your textbook.
☐ Set your learning goals. When you finish this exercise, you should be able to:
  ☐ use anatomical terms correctly.
  ☐ discuss the nature of an anatomical section.
  ☐ describe the basic plan of the human body.
  ☐ identify the major body cavities.
  ☐ list the major systems of the body, their principal organs, and their primary functions.
☐ Prepare your materials:
  ☐ dissectible human torso model (or comparable charts)
  ☐ models or figures showing different anatomical sections
  ☐ colored pencils or pens
  ☐ computer setup with human dissection program (optional)
☐ Read the directions for this exercise *carefully* before starting any procedure.

### *Study* TIPS

Learning and comprehending the material in this lab are important to your success in anatomy and physiology. These terms will be used extensively throughout the course. Use all of the available resources (textbook, colored illustrations, Brief Atlas, the *A&P Survival Guide*, flash cards, etc.) to review the terms. During this lab, use the tables and the illustrations, and with a lab partner work together in either labeling a torso chart or quizzing each other by identifying the various directional terms, planes and sections, body cavities, body regions, and body systems.

## A. ANATOMICAL POSITION

All terms describing the anatomy of organisms assume that the body is in the classic anatomical position. In this position, the body is in an erect (standing) posture with the arms at the sides and palms turned forward **(Figure 1-1)**. The head and feet are also pointing forward. The anatomical position is a reference position that gives meaning to the directional terms used to describe the body parts and regions.

☐ 1 Provide verbal instructions to your lab partner to position her/him in anatomical position.

> **Hint** ▶ The anatomical position of a four-legged animal, such as a rat, cat, or fetal pig, is standing on all four limbs, head facing forward. See **Figure 2-1, A** in Lab Exercise 2—Dissection: The Whole Body for an example. •

## B. ANATOMICAL DIRECTIONS

To locate structures within a body, you must use *directional terms*. Actually, you use these kinds of terms all the time—for example, *left*, *right*, *up*, *down*, *north*, and *south*.

☐ 1 Review the directional terms given in **Table 1-1**. Notice that they are grouped in relative pairs. Each member of a pair is the opposite, or complement, of the other member of the pair. For example, *right* is the opposite direction of *left*. To make the reading of anatomical figures a little easier, an *anatomical compass* is used throughout this lab manual, just as it is in the textbook. On many figures, you will notice a small compass rosette similar to those on geographical maps. Rather than being labeled N, S, E, and W, the anatomical compass rosette is labeled with abbreviated anatomical directions.

☐ 2 Review this list of directional terms and abbreviations:
  ☐ A = Anterior
  ☐ D = Distal
  ☐ I = Inferior
  ☐ L (opposite R) = Left
  ☐ L (opposite M) = Lateral
  ☐ M = Medial
  ☐ P (opposite A) = Posterior
  ☐ P (opposite D) = Proximal
  ☐ R = Right
  ☐ S = Superior

☐ 3 Review the examples of how the anatomical compass rosette is used in illustrations by looking at **Figure 1-2, A, B, F, G**. Test your knowledge of directions and the use of the anatomical compass rosette by labeling the rosettes given in **Figure 1-2, C, D, E, H- J**.

## C. PLANES AND SECTIONS

It is often useful to show a figure of a *sectioned* human body or organ. A "section" refers to a part cut along a plane. A plane is a geometrical concept referring to an imagined flat surface.

**TABLE  1-1    Directional Terms**

| DIRECTIONAL TERM | DEFINITION | EXAMPLE OF USAGE |
|---|---|---|
| Left | To the left of the body (not your left, the subject's) | The stomach is to the *left* of the liver. |
| Right | To the right of the body or structure being studied | The *right* kidney is damaged. |
| Lateral | Toward the side; away from the midsagittal plane | The eyes are *lateral* to the nose. |
| Medial | Toward the midsagittal plane; away from the side | The eyes are *medial* to the ears. |
| Anterior | Toward the front of the body | The nose is on the *anterior* of the head. |
| Posterior | Toward the back (rear) of the body | The heel is *posterior* to the toes. |
| Superior | Toward the top of the body | The shoulders are *superior* to the hips. |
| Inferior | Toward the bottom of the body | The stomach is *inferior* to the heart. |
| Dorsal | Along (or toward) the vertebral surface of the body | Her scar is along the *dorsal* surface. |
| Ventral | Along (toward) the belly surface of the body | The navel is on the *ventral* surface. |
| Caudad (caudal) | Toward the tail (four-legged animals) | The neck is *caudad* to the skull. |
| Cephalad | Toward the head (four-legged animals) | The neck is *cephalad* to the tail. |
| Proximal | Toward the trunk (describes relative position in a limb or other appendage) | The knee is *proximal* to the ankle. |
| Distal | Away from the trunk or point of attachment | The hand is *distal* to the elbow. |
| Visceral | Toward an internal organ; away from the outer wall (describes positions inside a body cavity) | This organ is covered with the *visceral* layer of the membrane. |
| Parietal | Toward the wall; away from internal structures | The abdominal cavity is lined with the *parietal* peritoneal membrane. |
| Deep | Toward the inside of a part; away from the surface | The thigh muscles are *deep* to the skin. |
| Superficial | Toward the surface of a part; away from the inside | The skin is a *superficial* organ. |
| Medullary | Refers to an inner region, or *medulla* | The *medullary* portion of the organ contains nerve tissue. |
| Cortical | Refers to an outer region, or *cortex* | The *cortical* area produces hormones. |
| Supine | Refers to lying face upward | The victim was found *supine* on the ground. |
| Prone | Refers to lying face downward | Examination of the spine can be performed with the patient in the *prone* position. |

The term *cross-section (c.s.)*, for example, refers to a part cut crosswise. A *longitudinal section (l.s.)* is a cut made lengthwise. These terms are useful only in limited circumstances because they do not really identify whether the cuts are made top to bottom, front to back, or side to side. The following three anatomical planes are used to describe sections of the body:

☐ 1 **Sagittal plane**—A sagittal plane extends from anterior to posterior and superior to inferior, dividing the body into left and right portions. A *midsagittal plane* refers to a sagittal plane that divides the body into exactly equal left and right portions.

☐ 2 **Frontal plane**—The frontal plane, also called a *coronal plane*, divides the body into anterior and posterior portions.

☐ 3 **Horizontal plane**—Also called a *transverse plane*, the horizontal plane divides the body into superior and inferior portions.

Using the models or figures provided, find at least three examples of each of the sections described (see **Figure 1-1**). For each, ask yourself what perspective the section gives that a section cut along a different plane does not give.

✏ *Coloring*
*Exercises:* **Planes and Directions**

Use colored pens or pencils to shade in both the figure and the labels. Each red numeral in the figure corresponds to a matching red numeral following the appropriate label.

## Planes
SAGITTAL₁
MIDSAGITTAL₂
FRONTAL₃
HORIZONTAL₄

## Directions
LATERAL₅
MEDIAL₆
ANTERIOR₇
POSTERIOR₈
SUPERIOR₉
INFERIOR₁₀
PROXIMAL₁₁
DISTAL₁₂

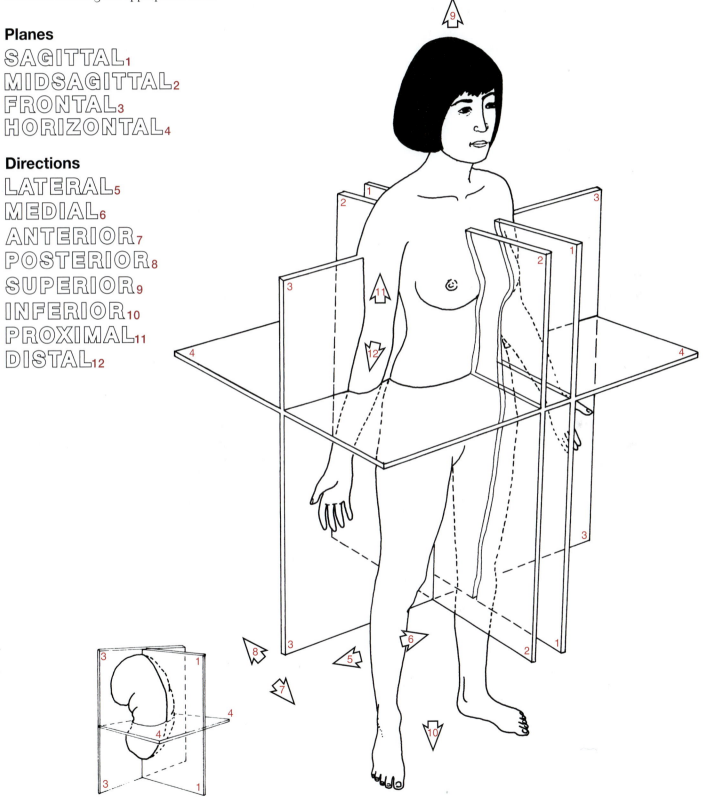

**FIGURE 1-1**

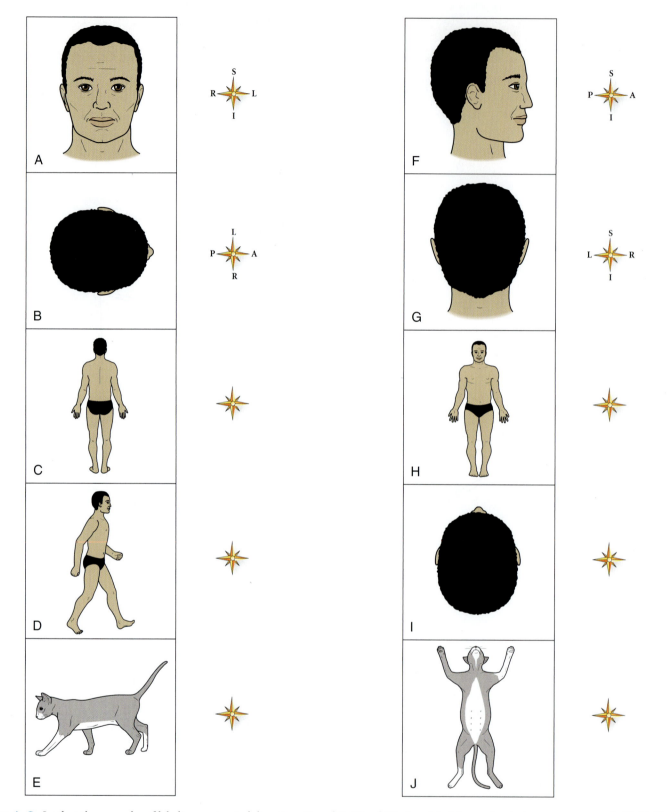

**FIGURE 1-2** Look at the examples of labeling anatomical directions (**A and B, F and G**); then label the anatomical compass rosettes in **C, D, E, H, I,** and **J** with the appropriate letters: **S, I, A, P, L,** and **R.**

# D. BODY CAVITIES AND REGIONS

The inside of the human body contains the viscera, or internal organs. The viscera are found in any of a number of cavities (spaces) within the body **(Figure 1-3)**. The two principal groups of body cavities are the **dorsal cavities** and the **ventral cavities**. Because these spaces are so large, they are sometimes subdivided into smaller units.

☐ 1 Using a dissectible torso model, find these divisions of the dorsal cavities (and the organs within):
  ☐ **Cranial cavity**—Within the skull. Organ: *brain*
  ☐ **Spinal cavity**—Within the vertebral column. Organ: *spinal cord*

☐ 2 Using the torso model, find these divisions and organs of the ventral cavities:
  ☐ **Thoracic cavity**—Within the rib cage
  ☐ **Pleural cavities**—Left one-third and right one-third of the thoracic cavity. Organ: *lung*

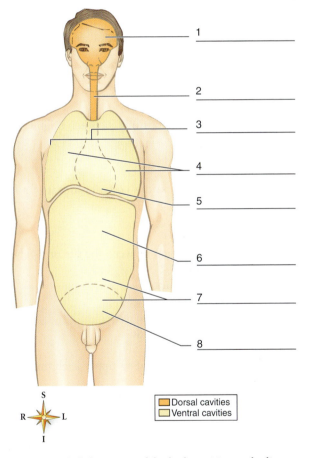

1 _____
2 _____
3 _____
4 _____
5 _____
6 _____
7 _____
8 _____

S
R ☆ L
I

☐ Dorsal cavities
☐ Ventral cavities

**FIGURE 1-3** Label the names of the body cavities on the lines provided and on the blanks in the Lab Report at the end of this exercise.

  ☐ **Mediastinum**—Middle one-third of thorax. Organs: *heart, trachea, esophagus*
  ☐ **Abdominopelvic cavity**—From the diaphragm to the bottom of the trunk
  ☐ **Abdominal cavity**—From the diaphragm to the rim of the pelvic bones. Organs: *stomach, liver, most of the intestines, pancreas, spleen, kidneys*
  ☐ **Pelvic cavity**—From the pelvic rim to the floor of the trunk. Organs: *portions of the intestines, ovaries, uterus, urinary bladder*

☐ 3 Because the abdominopelvic cavity is so large and contains so many different organs, it is often convenient to subdivide it into nine abdominopelvic regions **(Figure 1-4)**. The regions are bounded by a grid made by imagining two horizontal planes (one just below the ribs, the other just above the hip bones) and two sagittal planes (each just medial to a nipple). This arrangement forms a 3-D, tic-tac-toe grid in the abdominopelvic cavity. Label each of the nine regions on **Figure 1-4**, and then identify the approximate location of each region on a model of the human torso:
  ☐ **Right hypochondriac region**—Top right region (*hypochondriac* means "below [rib] cartilage")
  ☐ **Epigastric region**—Top middle region (*epigastric* means "near the stomach")
  ☐ **Left hypochondriac region**—Top left region
  ☐ **Right lumbar region**—Middle right region (*lumbar* refers to lumbar vertebrae in lower back)
  ☐ **Umbilical region**—Central region (*umbilical* refers to the umbilicus, or navel)
  ☐ **Left lumbar region**—Middle left region (*lumbar* refers to lumbar vertebrae in lower back)
  ☐ **Right iliac region**—Lower right region (*iliac* refers to ilium, the bowl-like part of the hip bone)
  ☐ **Hypogastric region**—Lower middle region (*hypogastric* means "below the stomach")
  ☐ **Left iliac region**—Lower left region

☐ 4 Physicians and other health professionals often use a simpler method to divide the abdominopelvic cavity. Using one horizontal line intersecting with one vertical line at the umbilicus, the abdominopelvic cavity can be easily segmented into four quadrants (see **Figure 1-4**). Identify the approximate locations of each of the four quadrants on a model of the human torso:
  ☐ **Right upper quadrant (RUQ)**
  ☐ **Left upper quadrant (LUQ)**
  ☐ **Right lower quadrant (RLQ)**
  ☐ **Left lower quadrant (LLQ)**

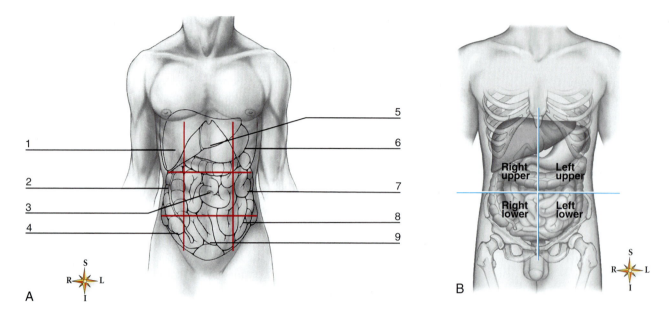

**FIGURE 1-4 A,** Label on the lines provided and on the blanks in the Lab Report the nine regions of the abdominopelvic cavity. **B,** Four quadrants of the abdominopelvic cavity.

## E. SURFACE REGIONS

There are hundreds of terms that describe specific locations on the surface of the human body. These names are useful not only for identifying surface features but also for the underlying muscles, bones, nerves, and blood vessels. In this activity, locate regions named by a few of the more common terms.

☐ 1 Locate the following surface regions on the anterior aspect of a human model or **Figure 1-5**:
- ☐ **Abdominal**—Area overlying the abdominal cavity
- ☐ **Antebrachial**—Forearm
- ☐ **Axillary**—Armpit
- ☐ **Brachial**—Upper arm
- ☐ **Buccal**—Cheek (side of mouth)
- ☐ **Carpal**—Wrist
- ☐ **Cervical**—Neck
- ☐ **Coxal**—Hip
- ☐ **Crural**—Anterior leg (shin)
- ☐ **Cubital**—Anterior of elbow
- ☐ **Femoral**—Thigh
- ☐ **Fibular (peroneal)**—Lateral leg
- ☐ **Mental**—Chin
- ☐ **Nasal**—Nose
- ☐ **Oral**—Mouth
- ☐ **Orbital**—Eye
- ☐ **Patellar**—Anterior knee joint

- ☐ **Pubic**—Lower front of trunk, between thighs
- ☐ **Tarsal**—Ankle
- ☐ **Thoracic**—Chest
- ☐ **Umbilical**—Navel

☐ 2 Identify these regions on the posterior aspect of a human model or **Figure 1-5**:
- ☐ **Calcaneal**—Heel
- ☐ **Cervical**—Neck
- ☐ **Flank**—Lateral region between the ribs and pelvis
- ☐ **Gluteal**—Buttocks
- ☐ **Lumbar**—Lower back
- ☐ **Occipital**—Posterior of head
- ☐ **Popliteal**—Posterior knee joint
- ☐ **Scapular**—Shoulder blade
- ☐ **Sural**—Calf
- ☐ **Thoracic**—Chest (including upper back through mid-back)

## F. BODY SYSTEMS

As you know, the human organism is composed of organ groups called *systems*. The organs of a system work together in an organized manner to accomplish the function(s) of the system. As an introduction to human body systems, study **Table 1-2**. Each of the systems will be discussed in more detail later in this course.

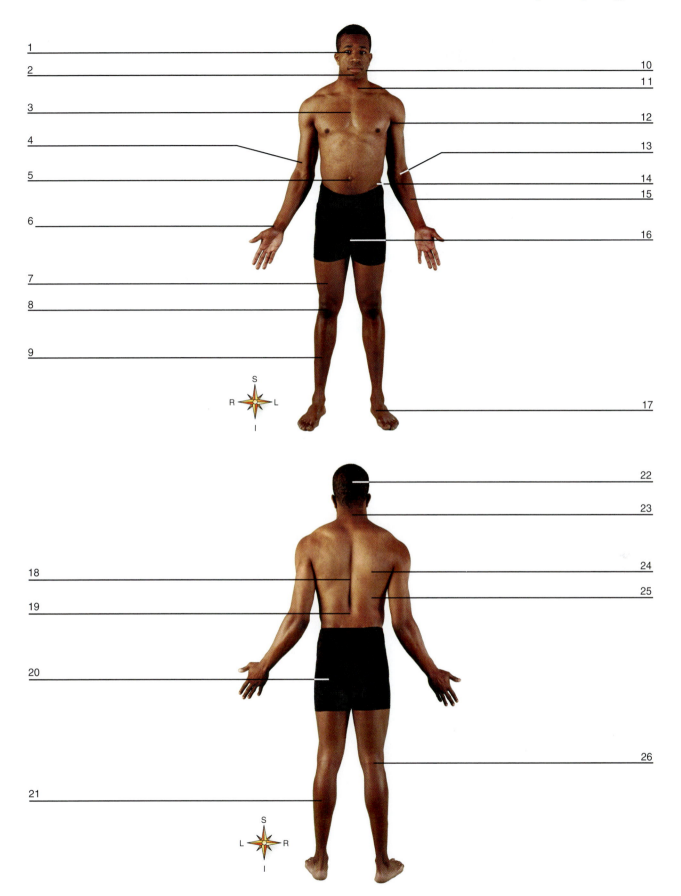

1 _____

2 _____

3 _____

4 _____

5 _____

6 _____

7 _____

8 _____

9 _____

10 _____

11 _____

12 _____

13 _____

14 _____

15 _____

16 _____

17 _____

18 _____

19 _____

20 _____

21 _____

22 _____

23 _____

24 _____

25 _____

26 _____

**FIGURE 1-5** Label these figures using the regional terms listed in activity E and on the blanks in the Lab Report at the end of this exercise.

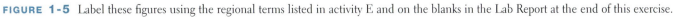

**T A B L E   1 - 2    Body Systems**

| ORGAN SYSTEM | PRINCIPAL ORGANS | PRIMARY FUNCTION(S) |
| --- | --- | --- |
| Integumentary | Skin | Protection, temperature regulation, sensation |
| Skeletal | Bones, ligaments | Support, protection, movement, mineral/fat storage, blood production |
| Muscular | Skeletal muscles, tendons | Movement, posture, heat production |
| Nervous | Brain, spinal cord, nerves, sensory organs | Control/regulation/coordination of other systems, sensation, memory |
| Endocrine | Pituitary gland, adrenals, pancreas, thyroid, parathyroids, other glands | Control/regulation of other systems |
| Cardiovascular | Heart, arteries, veins, capillaries | Exchange and transport of materials |
| Lymphatic | Lymph nodes, lymphatic vessels, spleen, thymus, tonsils | Immunity, fluid balance |
| Respiratory | Lungs, bronchial tree, trachea, larynx, nasal cavity | Gas exchange, acid-base balance |
| Digestive | Stomach, intestines, esophagus, liver, mouth, pancreas | Breakdown and absorption of nutrients, elimination of waste |
| Urinary | Kidneys, ureters, bladder, urethra | Excretion of waste, fluid and electrolyte balance, acid-base balance |
| Reproductive male | Testes, vas deferens, prostate, seminal vesicles, penis | Continuity of genetic information (reproduction) |
| Reproductive female | Ovaries, fallopian tubes, uterus, vagina | Reproduction, nurturing of offspring |

Name: _____ Date: _____ Section: _____

## LAB REPORT 1
# Organization of the Body

**Figure 1-3** (from p. 5)

1. _____
2. _____
3. _____
4. _____
5. _____
6. _____
7. _____
8. _____

**Figure 1-4** (from p. 6)

1. _____
2. _____
3. _____
4. _____
5. _____
6. _____
7. _____
8. _____
9. _____

**Multiple Choice**

1. _____
2. _____
3. _____
4. _____
5. _____
6. _____

**Figure 1-5** (from p. 7)

1. _____
2. _____
3. _____
4. _____

**Multiple Choice** (only one response is correct in each item)

1. An anatomist cuts a cadaver (preserved body) with a large saw in a way that divides the cadaver into equal left and right halves. The cut is along a _?_ plane.
   a. sagittal
   b. midsagittal
   c. frontal
   d. horizontal
   e. a and b are correct

2. In many skulls used for study in lab, the top of the skull can be removed so that inner features can be seen. Along which plane should one cut to open the top of a human study skull?
   a. sagittal
   b. coronal
   c. horizontal
   d. frontal
   e. b and d are correct

3. A surgeon makes an incision medially from the left axillary region, turning inferiorly at the midline and proceeding to the pubic region. The path of the cut can be mapped on the patient's chest as:
   a. ⌈  b. +  c. ∟  d. ¬  e. –  f. /

4. Which of these regions contains the spleen?
   a. epigastric
   b. hypogastric
   c. left hypochondriac
   d. right hypochondriac
   e. right lumbar

5. Soccer players often wear shin protectors, which shield the _?_ region of each leg.
   a. femoral
   b. popliteal
   c. cortical
   d. gluteal

6. Control and regulation of other systems are primary functions of the _?_.
   a. nervous system
   b. cardiovascular system
   c. endocrine system
   d. urinary system
   e. a and c are correct

5. _____

6. _____

7. _____

8. _____

9. _____

10. _____

11. _____

12. _____

13. _____

14. _____

15. _____

16. _____

17. _____

18. _____

19. _____

20. _____

21. _____

22. _____

23. _____

24. _____

25. _____

26. _____

**Fill-in**

1. _____

2. _____

3. _____

4. _____

5. _____

6. _____

7. _____

8. _____

9. _____

10. _____

11. _____

**Fill-in** (give the correct term for each item below; hint: remember to consider anatomical position as needed for determining directional terms)

1. The head is _?_ to the feet.

2. The liver is part of the _?_ system.

3. A leg amputation is likely to involve a _?_ cut, or section, through bone.

4. My lower back, or _?_, is sore.

5. The first finger is _?_ to the hand.

6. The popliteal vein is found in the _?_.

7. The heart is _?_ to the right lung.

8. The shoulder is _?_ to the elbow.

9. The skin is _?_ relative to the skeleton.

10. Adipose tissue is often just _?_ to the skin.

11. An occipital scar is on the back of the _?_.

**Sketch** (make a rough sketch of a human figure in the position indicated by the anatomical compass rosette and label)

# LAB EXERCISE 2
# Dissection: The Whole Body

This exercise may be used either as an introduction to vertebrate anatomy or as a synthesis activity to conclude the laboratory course. Succeeding exercises in this manual concentrate on organs or systems one by one. This exercise begins the study of the body with an overview of everything together in an exploration of the entire vertebrate body. Because the vertebrate body plan is similar among species, another lab animal specimen (e.g., fetal pig, rabbit) could be substituted.

 **Hint** In a short course, a well-preserved specimen can be used from time to time throughout your studies. If you will be looking at your dissected specimen for the next several months, make sure that it is kept in the appropriate container under conditions suggested by your instructor. •

## BEFORE YOU BEGIN

☐ Set your learning goals. When you finish this exercise, you should be able to:
    ☐ perform a whole-body dissection of a vertebrate animal.
    ☐ identify the major anatomical features of the vertebrate body in a dissected specimen.
☐ Prepare your materials:
    ☐ preserved (plain or double-injected) or fresh lab rat (or similar vertebrate)
    ☐ dissection tools and trays
    ☐ protective gear: gloves, safety eyewear, and apron
    ☐ mounted rat skeleton (optional)
    ☐ storage container (if preserved specimen is to be reused)
    ☐ computer setup with a program to simulate dissection of a human body (optional)
☐ Read the directions and safety tips for this exercise carefully before starting any procedure.

**🛈 safety first**
Observe the usual precautions when working with a preserved or fresh specimen. Heed the safety advice accompanying preservatives used with your specimen. Use protective gloves and safety eyewear while handling your specimen. Avoid injury with dissection tools. Dispose of your specimen as instructed. •

## Study TIPS

As you complete this dissection, refer to an illustration or atlas of the human body. Use the illustration or atlas to locate and compare the various external and internal structures with your preserved specimen. Because the vertebrate body plan among species is very similar, this may help you to understand the structure of the human body.

## A. EXTERNAL ASPECT

Examine the external aspect of your specimen.

☐ 1 Determine the anatomical orientation of the specimen. Which direction is anterior? Posterior? Which direction is ventral? Dorsal? Identify sagittal, transverse, and frontal planes in your specimen **(Figure 2-1, A)**.
☐ 2 Identify these externally visible features:
    ☐ **Pinna (auricle)**
    ☐ **External nares (nostrils)**
    ☐ **Vibrissae (whiskers)**
    ☐ **Incisors**
    ☐ **Integument**
    ☐ **Forelimbs**
    ☐ **Hindlimbs**
    ☐ **Thoracic region**
    ☐ **Abdominal region**
    ☐ **Nipples**
    ☐ **Anus**
    ☐ **Tail**
☐ 3 Determine the sex of your specimen by examining the external genitals:
    ☐ **Female**—Immediately anterior to the anus, on the ventral surface, is the **vulva** with an opening to the **vagina.** Anterior to the vulva is the **clitoris** with the **urethral** opening.
    ☐ **Male**—The **scrotum** containing the **testes** is immediately anterior to the anus, perhaps even hiding the anus from view. Near the anterior edge of the scrotum is the **penis** with its prepuce, or skinfold covering. Locate the opening of the **urethra** in the penis.

 **Hint** A complete set of anatomical sketches of the rat are found in the Anatomical Atlas of the Rat (see **Figures 2-1 to 2-7**). •

## B. SKIN, BONES, AND MUSCLE

☐ 1 Remove most of the skin from your specimen by following this procedure:

    ☐ Place the animal in the tray with its ventral surface facing you.

    ☐ Pull up on the skin over the sternum and puncture it with the tip of a scissors. Slide the bottom tip of the scissors into the **subcutaneous** area under the skin.

    ☐ Begin cutting along the lines indicated in **Figure 2-2.** Be careful not to cut into the skeletal muscles under the skin.

    ☐ With your forceps, pull the two flaps of skin over the neck away from the animal's body. Notice the **areolar tissue** under the skin that is pulled apart as you remove the skin. Pull the flaps of skin over the abdomen and over the **groin** area away in a similar fashion.

    ☐ Turn the animal over so the dorsum is facing you.

    ☐ If you made all the cuts properly, you should be able to pull the skin off the back in one complete piece. Sometimes it helps if you scrape at the loose connective tissue under the skin with your scalpel as you peel away the skin. The skinning process is difficult unless you have patience and proceed slowly.

☐ 2 Examine the skin, identifying these features:

    ☐ **Dermis**—The thick inner layer of the skin

    ☐ **Epidermis**—The thinner outer layer of the skin with *hair* (fur)

☐ 3 Explore the shape of the skinned rat body. How many **bones** of the rat's skeleton can you see or palpate (feel)? If you have a mounted rat skeleton available, identify as many of the bones of the skeleton as you can. If you become stumped, refer to **Figure 2-3, A,** in the rat atlas. Notice the similarity between the rat's skeletal plan and that of the human.

☐ 4 Observe the rat's musculature. Some of the external muscles of the torso can be separated from one another for easier viewing. Slide a probe into the loose connective tissue joining adjacent muscles and run the probe along their margins. Using **Figures 2-3** and **2-4** as guides, try to identify the major superficial muscles of the rat's body.

## C. CARDIOVASCULAR STRUCTURES

☐ 1 Open the **ventral body cavity** by cutting into its muscular wall in a manner similar to your earlier cut into the skin. Cut flaps in the neck, abdomen, and groin areas as you did with the skin, but do not cut all the way around to the back. Be careful not to damage any **visceral organs** with your scissors as you cut. Fold back the flaps, and either anchor them with pins or remove them.

☐ 2 Locate the **heart** near the middle of the **thoracic cavity** in the **mediastinum.** Can you identify the four chambers?

☐ 3 If you have a double-injected preserved specimen, the **arteries** are filled with red latex and the **veins** are filled with blue latex. If not, the arteries can usually be distinguished from veins because they are stiffer and lighter in color than

veins. Locate the **aorta,** the large artery leaving the heart and arching posteriorly. Trace the branches of the aorta, naming them if you can. Use **Figure 2-5, A,** if you need help.

☐ 4 Locate the **anterior vena cava** where it drains into the heart. This is analogous to the superior vena cava in the human. Follow its tributary veins and identify them with the help of **Figure 2-5, B.** Locate the **posterior vena cava** and trace its tributaries.

> **Hint** ▶ Once you have cut into the ventral body cavity, you may be tempted to cut and remove organs. It is important that you keep everything as intact as possible. You may pull organs to one side or another to view deeper structures, but avoid making cuts. •

## D. VISCERA

The viscera, or major internal organs, can be seen within the ventral body cavity. Use **Figures 2-6** and **2-7** to guide you in locating the following:

☐ 1 Locate some of these features of the lower **respiratory system:**

    ☐ Larynx

    ☐ Trachea

    ☐ Primary bronchi

    ☐ Lungs (Can you distinguish the parietal and visceral pleurae?)

    ☐ Diaphragm

☐ 2 Locate these structures of the **digestive system:**

    ☐ Submandibular salivary glands

    ☐ Esophagus

    ☐ Stomach

    ☐ Liver

    ☐ Pancreas

    ☐ Small intestine

    ☐ Mesentery

    ☐ Cecum (Very large in the rat and does not have an appendix as does the human cecum.)

    ☐ Colon

☐ 3 Locate these **lymphatic** organs:

    ☐ Spleen

    ☐ Thymus (May be small or not visible in older animals.)

☐ 4 Locate these features of the **urinary system:**

    ☐ Kidney

    ☐ Ureter

    ☐ Urinary bladder

    ☐ Urethra

☐ 5 Try to locate these **endocrine glands** in your specimen:

    ☐ Thyroid gland

    ☐ Thymus gland

    ☐ Pancreas

    ☐ Adrenal glands

    ☐ Testes

    ☐ Ovaries

☐ 6 Identify these structures associated with the **male reproductive system:**
- ☐ Testes
- ☐ Epididymis
- ☐ Ductus (vas) deferens
- ☐ Seminal vesicle
- ☐ Prostate gland
- ☐ Penis

☐ 7 Find these **female reproductive system** structures:
- ☐ Ovaries
- ☐ Uterine tubes
- ☐ Uterus (Notice that the rat uterus has a Y shape, with a right and left uterine horn.)
- ☐ Vagina

☐ 8 If your specimen is a pregnant female, open the uterus *very carefully* and identify these structures:
- ☐ **Amniotic sac**—This is a fluid-filled cushion for each embryo.
- ☐ **Embryo**—Examine the structure of the embryo. Do you think that it is in a late stage or an early stage of development?
- ☐ **Umbilical cord**—Can you distinguish the umbilical vessels?
- ☐ **Placenta**—It is attached to the uterine lining.

**Hint** Unless your lab group has both male and female specimens, you may want to temporarily trade specimens with a group that has a rat of a different sex than yours. By doing so, you will be able to find the features of both reproductive systems. •

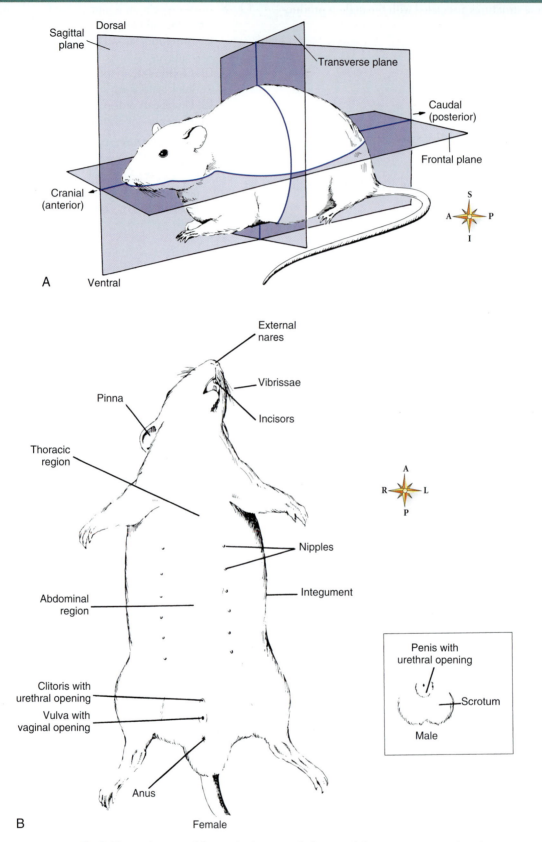

**FIGURE 2-1** External aspect of the rat. **A,** Anatomical planes and directions. **B,** Ventral surface.

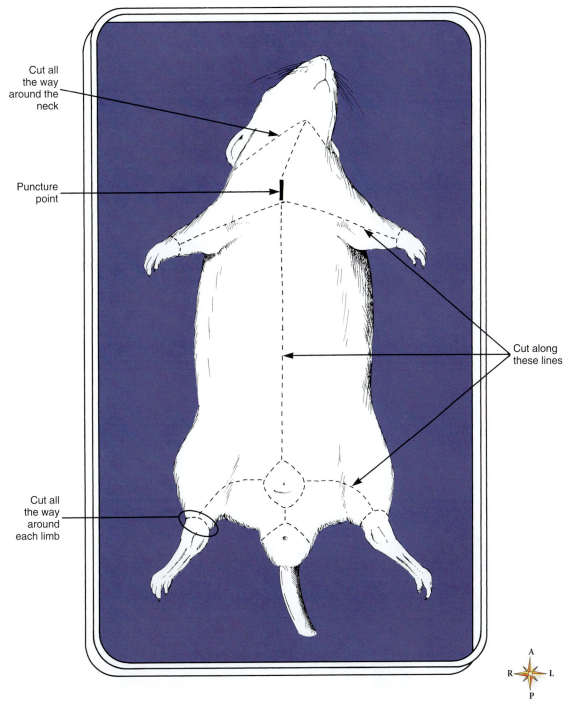

Cut all
the way
around the
neck

Puncture
point

Cut along
these lines

Cut all
the way
around each limb

**FIGURE 2-2** Directions for cutting the skin of the rat.

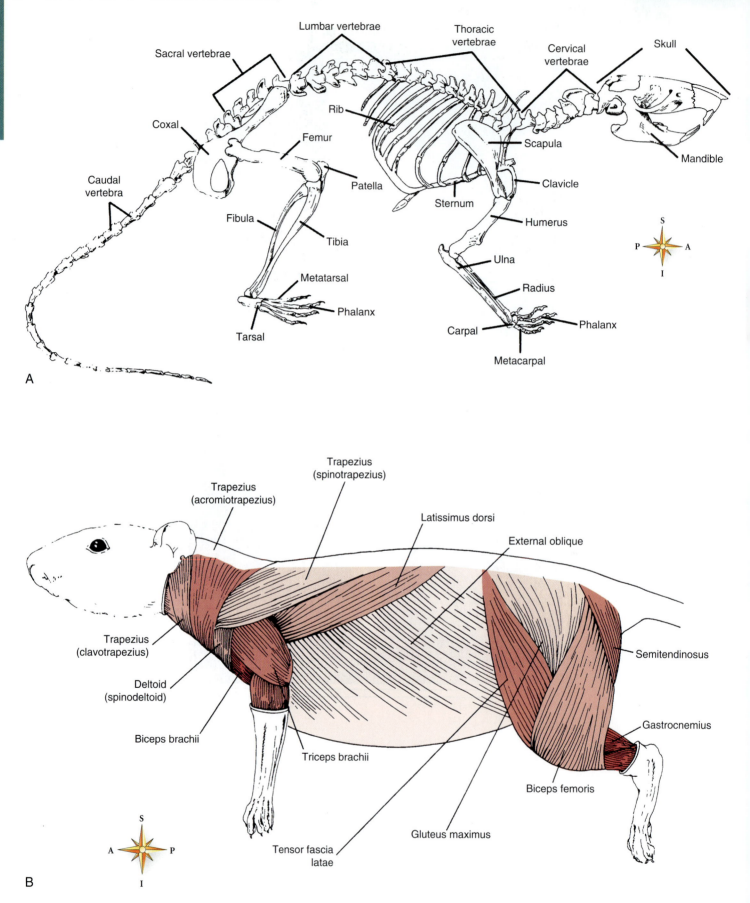

**FIGURE 2-3** Lateral views of the rat. **A,** The skeleton. **B,** Superficial skeletal muscles.

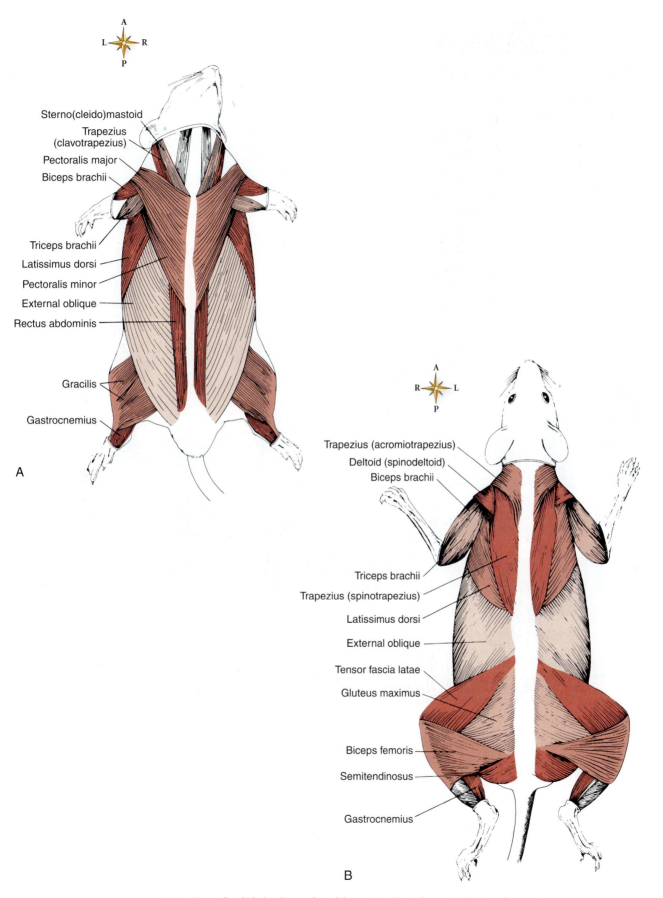

**FIGURE 2-4** Superficial skeletal muscles of the rat. **A,** Ventral aspect. **B,** Dorsal aspect.

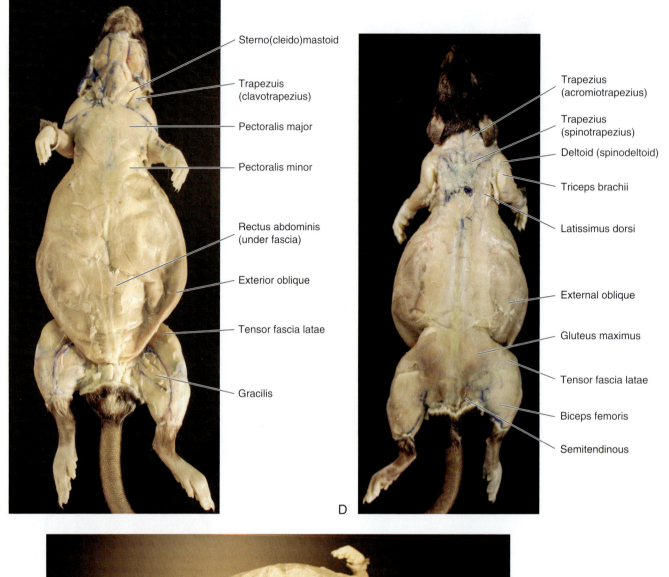

C — Sterno(cleido)mastoid
Trapezuis (clavotrapezius)
Pectoralis major
Pectoralis minor
Rectus abdominis (under fascia)
Exterior oblique
Tensor fascia latae
Gracilis

D — Trapezius (acromiotrapezius)
Trapezius (spinotrapezius)
Deltoid (spinodeltoid)
Triceps brachii
Latissimus dorsi
External oblique
Gluteus maximus
Tensor fascia latae
Biceps femoris
Semitendinous

E — Gastrocnemius   Gracilis   Tensor fascia latae   Exterior oblique   Pectoralis minor   Pectoralis major   Biceps brachii   Trapezuis (clavotrapezius)   Sterno (cleido) mastoid

**FIGURE 2-4, cont'd C,** Ventral aspect dissected. **D,** Dorsal aspect dissected. **E,** Lateral aspect dissected.

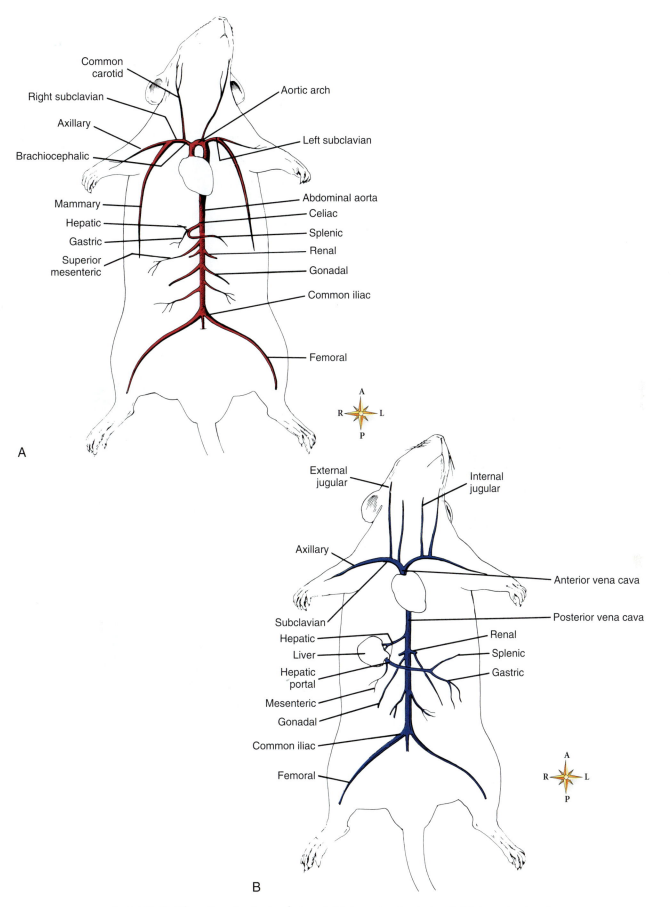

**FIGURE 2-5** The rat's systemic circulation. **A,** Major systemic arteries. **B,** Major systemic veins.

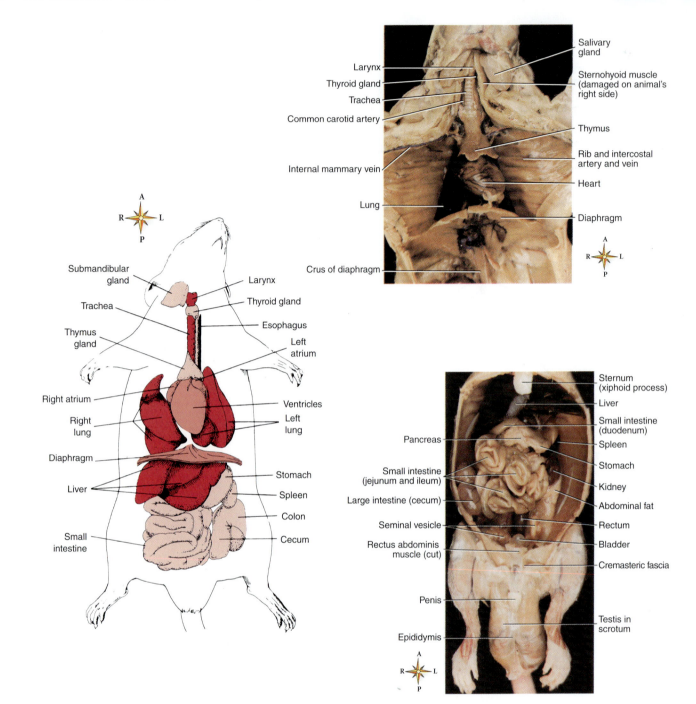

**FIGURE 2-6**  Ventral body cavity of the rat.

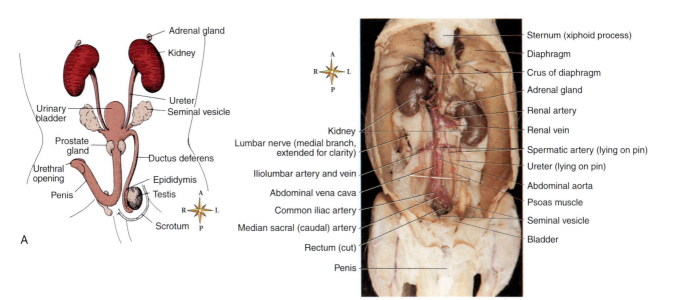

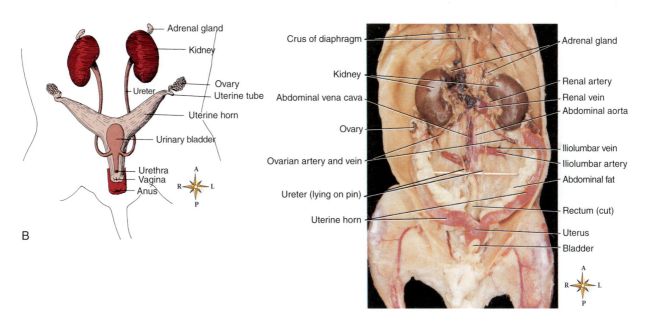

FIGURE 2-7    Pelvic body cavity of the rat. **A**, Male. **B**, Female.

Name: _____ Date: _____ Section: _____

# LAB REPORT 2
# Dissection: The Whole Body

## Rat Dissection Checklist

☐ **External aspect**
  ☐ pinna (auricle)
  ☐ external naris (nostril)
  ☐ vibrissae (whiskers)
  ☐ incisors
  ☐ integument
  ☐ forelimbs
  ☐ hindlimbs
  ☐ thoracic region
  ☐ abdominal region
  ☐ nipples
  ☐ anus
  ☐ tail
  ☐ sex: female
    ☐ vulva
    ☐ vagina
    ☐ clitoris
    ☐ urethral opening
  ☐ sex: male
    ☐ scrotum (testes within)
    ☐ penis
    ☐ prepuce (foreskin)
    ☐ urethral opening
☐ **Skin, bones, and muscles**
  ☐ subcutaneous areolar tissue
  ☐ dermis
  ☐ epidermis, fur
  ☐ axial skeleton
    ☐ cranial bones
    ☐ facial bones
    ☐ cervical vertebrae
    ☐ thoracic vertebrae
    ☐ lumbar vertebrae
    ☐ sacral vertebrae
    ☐ caudal vertebrae
    ☐ ribs
    ☐ sternum
  ☐ appendicular skeleton
    ☐ scapula
    ☐ clavicle
    ☐ humerus
    ☐ radius
    ☐ ulna
    ☐ carpal bones
    ☐ metacarpal bones
    ☐ phalanges of the forelimb
    ☐ coxal bone
    ☐ femur
    ☐ patella

☐ tibia
☐ fibula
☐ tarsal bones
☐ metatarsal bones
☐ phalanges of the hindlimb
☐ head, neck, and shoulder muscles
  ☐ sternomastoid (human: sternocleidomastoid)
  ☐ clavotrapezius (human: trapezius)
  ☐ acromiotrapezius (human: trapezius)
  ☐ spinotrapezius (human: trapezius)
  ☐ spinodeltoid (human: deltoid)
  ☐ biceps brachii
  ☐ triceps brachii
  ☐ latissimus dorsi
☐ abdominal and hindlimb muscles
  ☐ external oblique
  ☐ rectus abdominis
  ☐ tensor fascia latae
  ☐ gluteus maximus
  ☐ biceps femoris
  ☐ semitendinosus
  ☐ gastrocnemius
  ☐ gracilis
☐ **Cardiovascular structures**
  ☐ heart
    ☐ left and right atria
    ☐ left and right ventricles
  ☐ aorta
  ☐ major aortic branches
    ☐ common carotid arteries
    ☐ brachiocephalic artery
    ☐ left subclavian artery
  ☐ celiac artery (trunk)
  ☐ superior mesenteric artery
  ☐ renal arteries
  ☐ gonadal arteries
  ☐ common iliac arteries
  ☐ anterior vena cava (human: superior vena cava)
  ☐ subclavian veins
  ☐ posterior vena cava (human: inferior vena cava)
  ☐ hepatic vein
  ☐ renal veins
  ☐ common iliac veins
  ☐ hepatic portal vein
☐ **Viscera**
  ☐ respiratory structures
    ☐ larynx
    ☐ trachea
    ☐ primary bronchi
    ☐ lungs

- ☐ pleurae
- ☐ diaphragm
- ☐ digestive structures
  - ☐ submandibular salivary glands
  - ☐ esophagus
  - ☐ stomach
  - ☐ liver
  - ☐ pancreas
  - ☐ small intestine
  - ☐ mesentery
  - ☐ cecum
  - ☐ colon
- ☐ lymphatic organs
  - ☐ spleen
  - ☐ thymus (if visible)
- ☐ urinary structures
  - ☐ kidney
  - ☐ ureter
  - ☐ urinary bladder
  - ☐ urethra
- ☐ endocrine glands

- ☐ thyroid gland
- ☐ thymus gland
- ☐ pancreas
- ☐ adrenal glands
- ☐ male reproductive organs
  - ☐ testes
  - ☐ epididymis
  - ☐ ductus (vas) deferens
  - ☐ seminal vesicles
  - ☐ prostate gland
  - ☐ penis
- ☐ female reproductive organs
  - ☐ ovaries
  - ☐ uterine tubes (oviducts)
  - ☐ uterus
  - ☐ vagina
- ☐ pregnant uterus
  - ☐ amniotic sac
  - ☐ embryo (or fetus)
  - ☐ umbilical cord
  - ☐ placenta

# LAB EXERCISE 3
# The Microscope

In the seventeenth century, the amateur Dutch scientist Anton van Leeuwenhoek used one of the first microscopes to discover a whole new world of living organisms. Using a single lens, or **simple microscope**, he observed tiny organisms in pond water and scrapings from his own teeth. Robert Hooke, an English scientist, discovered that larger organisms had small microscopic subunits that he called boxes or *cells*. Ever since this early era of discovery, biological microscopy has been essential in the study of living organisms.

The microscopes used in this course are **compound microscopes**, made of a set of lenses. They are more powerful and more complex than those used by van Leeuwenhoek and Hooke. This exercise introduces you to the use and care of the standard compound microscope.

## BEFORE YOU BEGIN

- ☐ Read the appropriate chapter in your textbook.
- ☐ Set your learning goals. When you finish this exercise, you should be able to:
  - ☐ identify each major part of a compound light microscope and describe its function
  - ☐ determine total magnification at different settings
  - ☐ use a microscope to observe prepared specimens
  - ☐ prepare a wet-mount slide
  - ☐ stain microscopic specimens
  - ☐ record microscopic observations accurately
- ☐ Prepare your materials:
  - ☐ compound light microscope
  - ☐ prepared microslides: *newsprint "e," three colored threads*
  - ☐ clean slides
  - ☐ coverslips
  - ☐ paper wipes
  - ☐ methylene blue stain
  - ☐ flat toothpicks
- ☐ Read the directions and safety tips for this exercise *carefully* before starting any procedure.

### *Study* TIPS

To become proficient in the use of the microscope, take advantage of this lab activity and any "open labs" that may be offered. Obviously, the more you use the microscope, the better you will become in focusing and examining specimens. With your instructor's permission, you may want to extend this lab activity by examining other specimens (hair, cloth fibers, and so on).

Proper identification of the various parts of the microscope is important to its proper use and care. Obtain an unlabeled copy of a microscope photo. Your instructor may have an unlabeled copy of a microscope photo, or you may want to photocopy the photo of the microscope **(Figure 3-1)** and delete the labels. You can also use your cell phone to photograph the microscope and later print the photo and label it. Quiz yourself by taking this blank photo and labeling the parts of the microscope.

## A. PARTS OF THE MICROSCOPE

- ☐ 1 Obtain a compound light microscope and identify each of the structures described in the following steps.

> **Hint** Because not all microscope models are alike, some of the structures described may be different on your scope. Refer to **Figure 3-1** for help in this exercise. •

- ☐ 2 Each lens closest to your eye is an **ocular** lens, or eyepiece. The ocular lens magnifies an image by the factor indicated on the ocular's barrel, usually 10×. If the factor is 10×, the image is magnified 10 times. If the factor is 5×, the image is magnified 5 times.
- ☐ 3 The **body tube** holds the ocular in place. Although called a *tube*, it may be more like a box in some models.
- ☐ 4 At the bottom of the body tube is the **revolving nosepiece**. A turretlike circular mechanism rotates so that different lenses can be selected. *Always* rotate the nosepiece by holding the outside of the revolving disk; *never* push on the lens barrels.
- ☐ 5 Each of the lens sets attached to the revolving nosepiece is an **objective lens**. As with the ocular lens, each objective lens is marked with its magnification factor. Microscopes may have any or all of the following objective lenses:
  - ☐ 4× (scanning objective) is used for initial location of the specimen. Some scopes have 3.2× or 3.5× scanning objective lenses. List the factor on your scope's scanning objective lens here. _____
  - ☐ 10× (low-power objective) may also be used for initial location of the specimen. It is also used for observing specimens that do not need greater magnification.

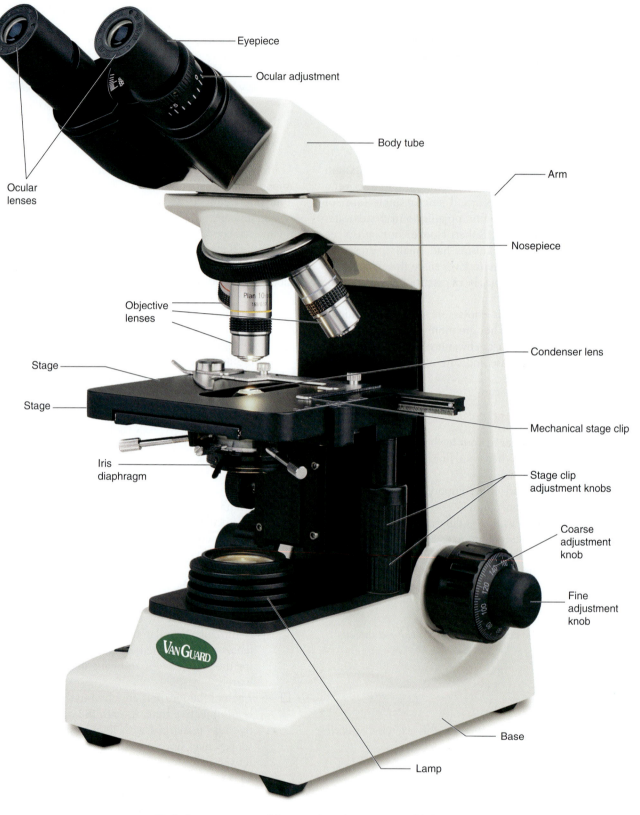

Eyepiece

Ocular adjustment

Body tube

Arm

Nosepiece

Ocular lenses

Objective lenses

Condenser lens

Stage

Stage

Mechanical stage clip

Iris diaphragm

Stage clip adjustment knobs

Coarse adjustment knob

Fine adjustment knob

Base

Lamp

**FIGURE 3-1**  Important parts of the typical modern compound light microscope.

☐ 40× (high-dry objective) is used for specimens requiring greater magnification. This objective lens is called *dry lens* because it does not require the use of oil, as other high-power objectives do. Some scopes have 43× or 45× high-dry objective lenses. List the factor on your scope's high-dry objective lens here. _____

☐ 100× (high-oil objective) is used for magnification of extremely small specimens, such as bacterial cells. It must be immersed in oil, so it is called the *high-oil* objective lens.

☐ In this manual, *low-power lens* will refer to use of the 10× objective lens, and *high-power lens* will refer to use of the 40× objective lens.

☐ 6 **Total magnification** is determined by multiplying the power of the ocular lens by the power of the objective lens in use. Thus, when using a 10× ocular lens and the low-power objective lens, total magnification is 100× (10 × 10 = 100). List all ocular and objective combinations, and then determine all the total magnifications possible on your scope and record them in the Lab Report at the end of this exercise.

☐ 7 The specimen is usually mounted on a glass or plastic **microscope slide** that rests on the stage, a platform just below the objective lens. The stage has a hole so that light can pass through the specimen from below. If the stage has an adjustable bracket that moves the slide around mechanically, the stage is called a **mechanical stage.** If not, the slide is held by **stage clips** and must be moved by hand.

☐ 8 Below the stage is a high-intensity **lamp**. Light rays from the lamp travel through a hole in the stage, through the specimen mounted on a slide, and then through the objective and ocular, to the eye.

☐ 9 A **condenser**, a lens that concentrates light, may be found between the lamp and the stage.

☐ 10 Sometimes the light from the lamp is too strong to see the specimen clearly. The light level may be reduced by adjusting the lamp intensity (if possible). Light intensity may also be adjusted by adjusting the **diaphragm** just below the stage (Figure 3-2). A *disk diaphragm* is a rotating disk with holes of different diameters. An *iris diaphragm* is made of overlapping slivers of metal in a pattern resembling the iris flower, as does the iris of the human eye. Like the iris of the eye, the iris diaphragm can dilate or constrict its opening. You can change the amount of light passing to the specimen either by rotating the edge of the disk diaphragm or by rotating the lever projecting from the iris diaphragm.

☐ 11 The entire upper assembly of the microscope is held in an upright position by a bar called the **arm**. The scope is supported by a square or horseshoe-shaped **base**. The arm may be connected to the base by a **pivot**, which allows the upper assembly to move into a more comfortable viewing position.

☐ 12 The **coarse-focus knobs** and **fine-focus knobs** are on the arm. These knobs adjust the distance between the stage and objective lens, thus focusing an image of the specimen. The fine-focus knob changes the distance very little, whereas the coarse-focus knob changes the distance greatly.

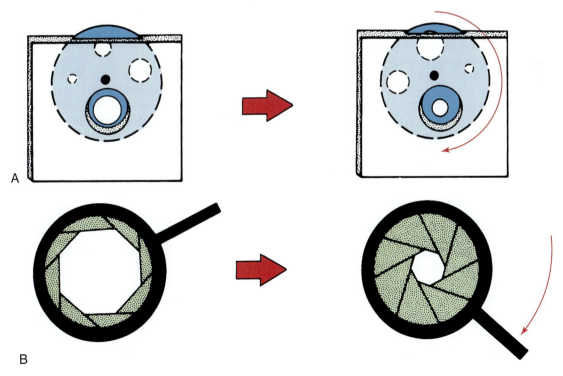

FIGURE 3-2 **A,** Disk diaphragm set at a large opening, then at a smaller opening. **B,** Iris diaphragm open, then partially closed.

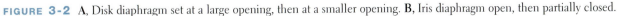

# B. USING THE MICROSCOPE

Skill in using the microscope is necessary for many of the exercises in this lab manual. Fortunately, learning to use the microscope is both easy and fun. Before beginning, acquaint yourself with these basic rules:

☐ Always carry the scope with two hands: one under the base and the other grasping the arm. Carry it in an upright position.

☐ Unwind the lamp cord carefully. Avoid damaging the parts around which it is coiled. Plug the cord into an outlet in a safe manner.

> **!** *safety first*
>
> Be careful when plugging the power cord into the outlet. Always plug it into a receptacle at your station, taking care not to string it over a chair, across an aisle, or in any other dangerous position. Before you turn on the microscope, make sure that the light intensity dial is on the lowest setting. If the light intensity dial is on a high setting, the sudden voltage surge may burn out the lamp.    •

☐ Make sure the stage and objective are at the farthest distance apart and that the *lowest-power objective is in position.* Start each new observation at low power.

Proceed with the following steps:

☐ 1  Obtain a microscope slide with a newsprint letter *"e"* mounted on it. This is your practice specimen.

☐ 2  Place your slide carefully on the stage and secure it with stage clips or the brackets of the mechanical stage. Move the slide so that the *"e"* is centered in the stage's hole.

☐ 3  While looking from the side, use the coarse-focus knob to move the objective lens (still on low power) as close as possible to the slide.

☐ 4  Look through the ocular lenses and use the coarse-focus knob to slowly move the objective lens and slide apart. When the image becomes clear, switch to the fine-focus knob to make the image even sharper. To avoid damaging the scope and slide, *never move the objective lens and stage toward each other while looking through the ocular lens.*

> **Hint**  To avoid eyestrain and a possible headache, always keep both eyes open when viewing a specimen. With binocular scopes this is easy once you have adjusted the distance between ocular lenses. However, it may take some practice if you are using a monocular scope. If you have trouble keeping both eyes open, try covering the unused eye with your hand.    •

☐ 5  Adjust the light intensity, using the lamp controller or diaphragm, until the detail of the image is at its clearest.

☐ 6  Sketch the entire field of view in the space provided in the Lab Report. You'll see that the *"e"* is upside down and backward. This inversion occurs with all specimens.

> **Hint**  Here are some rules for recording microscopic observations properly:
> - Label each sketch with the name of the specimen and what type of section it is.
> - Label each sketch with the total magnification.
> - Label as many parts of the sketch as you can. Labels should be orderly, never crossing lines with each other.
> - Sketches should be done with care, not with haste.

☐ 7  Center the part of the specimen that you wish to see more clearly and switch to a higher-power objective lens. Most scopes are **parfocal,** which means that the image remains focused when you change objective lenses. However, some minor adjustment with the fine focus is usually necessary. *Never adjust the coarse focus when using the high-power lens.*

☐ 8  Sketch the entire field that you see using the high-power (40× to 45×) objective lens.

☐ 9  When you remove the slide, make sure that the low-power objective lens is in viewing position and that the objective and stage are as far apart as possible.

☐ 10  Practice focusing at different depths by using a prepared slide of three crossed strands of colored thread. The strands (one red, one yellow, and one blue) cross at the same point. Determine which is on top, which is in the middle, and which is on the bottom. Report your results in the Lab Report.

☐ 11  When you have finished for the day, return the scope to its original configuration. The low-power objective lens should be in position. Make sure the slide has been removed.

> **Hint**  If you can't see an image in your scope:
> - Make sure the lamp is on.
> - Make sure the diaphragm is open. The amount of light should be set low on low power and high on high power.
> - Check for obstructions in the light path.
> - Make sure the objective is seated properly.
> - Check to see if the specimen is centered.
> - Clean the lenses with lens paper and lens solution (not facial tissues or lab wipes).
>
> If the image seems to fade in and out:
> - Watch to see if the body tube or stage is shifting or dropping.
> - Make sure the scope doesn't require a rubber eyecup for proper viewing.
> - Make sure the specimen is not in a medium that obstructs viewing.
> - Check the lamp or cord for short circuits.
>
> If the image doesn't look like the figure in the book:
> - Get a different book (it may have a better figure).
> - Use your imagination.
> - Join the club (nobody else's specimen is identical to that in the book either).

## C. PREPARING MICROSCOPIC SPECIMENS

We will often use prepared slides of human and other tissues in this course. However, occasionally we may be making our own specimens.

A **wet-mount** slide is a slide on which a wet specimen is placed, then covered with a **coverslip**. **Stains** are used to make a specimen, or some of its parts, more visible. Some require special techniques, but most stains can simply be added to the specimen and viewed. The usual method for preparing a stained wet-mount slide is described in **Figure 3-3** and in the following steps:

☐ 1 Obtain some skin cells by scraping the inner surface of your cheek with a clean, flat toothpick.

☐ 2 Wipe the scrapings on a clean microscope slide and put a small drop of *methylene blue* stain directly on the smear.

> **safety first**
>
> Use only fresh toothpicks, and dispose of used toothpicks immediately by placing them in the waste container indicated by your instructor. Take precautions to avoid the spread of disease through saliva. •

☐ 3 Place one edge of a coverslip on the slide next to the specimen, and then let it drop slowly on the specimen. This method prevents forming air bubbles.

☐ 4 Absorb any excess fluid around the edges of the coverslip with the edge of a paper wipe.

☐ 5 Locate some cells with the low-power objective lens, shift to the high-power lens, and sketch your observations in the Lab Report.

> **landmark characteristics**
>
> You are looking for scattered epithelial cells. Each cell will appear as a lightly stained, flat polygon or circle with a dark center. The dark center is the nucleus of the cell. Some cells may be separate, whereas others are clumped together.
>
> You may see a variety of other things in your specimen: bacteria, vegetable or meat fibers, and other matter. Don't worry; this is normal. Distinct dark circles with hollow centers that may be in motion are air bubbles. •

> **Hint** ▶ If your instructor wants you to learn how to use an oil-immersion objective lens, follow these directions. After focusing on the high-dry lens, rotate the nosepiece *halfway* to the high-oil objective lens. Put a drop of immersion oil on the slide, and then rotate the high-oil objective lens into the oil drop. You may need to increase the lighting. When finished, clean the objective lens and other surfaces as your instructor directs. •

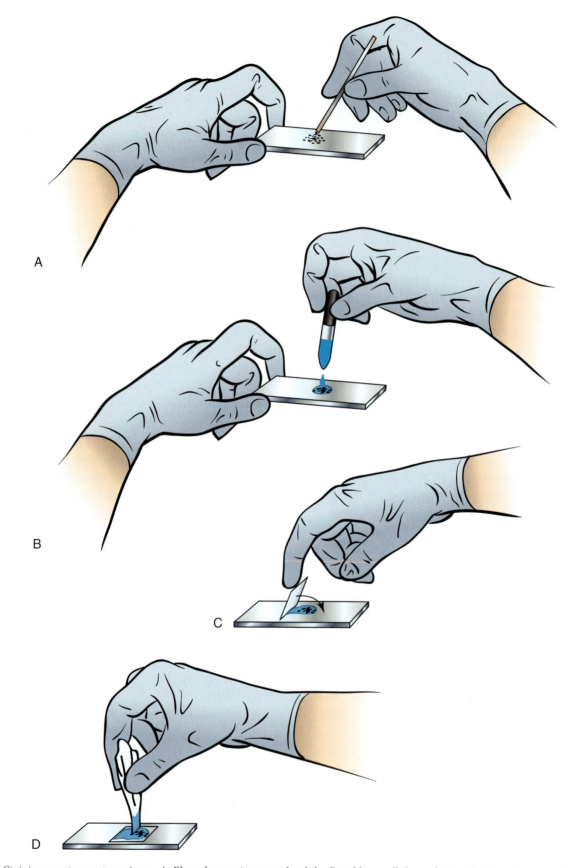

**FIGURE 3-3** Staining a wet-mount specimen. **A,** Place the specimen on the slide. **B,** Add a small drop of stain. **C,** Drop the coverslip slowly to prevent air bubbles. **D,** Absorb excess fluid with a paper wipe.

Name: _____ Date: _____ Section: _____

## LAB REPORT 3
# The Microscope

Does the "e" in the newsprint appear to be oriented the same way it is on the stage? Explain your observations.

*Specimen: newsprint "e"*
Total Magnification: _____

Is the *diameter of field* the same when observing the specimen under high power and low power?

Which of the three colored threads in the prepared slide is on the bottom?

Which thread is on the top?

*Specimen: newsprint "e"*
Total Magnification: _____

*Specimen: human cheek cells*
Total Magnification: _____

What is the darkly stained region of each cell?

What advantage is gained by staining this specimen?

List all the possible ocular-objective lens combinations on your microscope and calculate the total magnification for each.

| POWER OF OCULAR | POWER OF OBJECTIVE | TOTAL MAGNIFICATION |
|---|---|---|
|  |  |  |
|  |  |  |
|  |  |  |

## Fill-in

1. _____
2. _____
3. _____
4. _____
5. _____
6. _____
7. _____
8. _____

## Multiple Choice

1. _____
2. _____
3. _____
4. _____
5. _____
6. _____
7. _____

## Fill-in

1. A _?_ is a glass or plastic rectangle on which specimens are mounted.
2. A _?_ is at the base of the body tube, allowing the selection of different objective lenses.
3. Below the stage is a high-intensity _?_ or other light source.
4. Different lens sets, or _?_, are attached to the revolving nosepiece.
5. If the stage has a bracket that moves the slide, it is called a(n) _?_ stage.
6. The _?_ allows the user to limit the amount of light from a steady light source.
7. The set of lenses closest to the viewer's eye is the _?_.
8. The _?_ is a lens under the stage that concentrates light.

## Multiple Choice (choose the one best response)

1. A microscope _?_ the image of a specimen.
   a. magnifies
   b. reverses
   c. reduces
   d. erases
   e. a and b are correct
2. The use of _?_ may make the specimen (or some of its parts) more visible.
   a. stain
   b. the diaphragm
   c. fine focus
   d. all of the above
   e. none of the above
3. If you select a 40× objective lens to use with a 5× ocular lens, the total magnification would be:
   a. 200×
   b. 45×
   c. 400×
   d. 405×
   e. none of the above
4. When you move a specimen to the right, its image appears to
   a. move to the right
   b. move to the left
   c. remain stationary
5. Which is the correct path of light in a compound microscope?
   a. lamp, objective lens, diaphragm, specimen, ocular lens
   b. lamp, diaphragm, specimen, objective lens, ocular lens
   c. diaphragm, lamp, specimen, ocular lens, objective lens
   d. lamp, diaphragm, specimen, ocular lens, objective lens
6. If your scope is not working properly, consult
   a. the lab instructor
   b. your lab partner
   c. the owner's manual
   d. the manufacturer
7. When carrying your microscope, use _?_ hands.
   a. one
   b. two

# LAB EXERCISE 4
# Cell Anatomy

The cell is often considered the basic unit of living organisms. Actually, cells are only one of several *levels of organization* in the human organism:

ATOMS
MOLECULES
ORGANELLES
CELLS
TISSUES
ORGANS
ORGAN SYSTEMS
ORGANISM

Thus, cells are parts of larger units (tissues and organs) and are composed of smaller units (organelles).

For now, we will concentrate on the structure and function of the cell and its organelles. We will study some of the other levels of organization in later units.

## BEFORE YOU BEGIN

☐ Read the appropriate chapter in your textbook.
☐ Set your learning goals. When you finish this exercise, you should be able to:
  ☐ place the cell and its organelles within the scheme of organizational levels
  ☐ identify the major organelles on a model, chart, specimen, or micrograph
  ☐ describe primary function(s) of typical organelles
  ☐ discuss the anatomical relationship among various cell parts
☐ Prepare your materials:
  ☐ cell model or chart
  ☐ compound light microscope
  ☐ clean slides
  ☐ coverslips
  ☐ paper wipes
  ☐ methylene blue stain
  ☐ colored pencils or pens
  ☐ prepared slides (instructor's option)
☐ Read the directions and safety tips for this exercise *carefully* before starting any procedure.

## *Study* TIPS

When studying the structure and function of the cell, it might be helpful to make an analogy between a cell and a factory. Factories have various departments that work together to manufacture a specific product(s). Cells have specialized departments (organelles) that also work together. For instance, the cell's nucleus can be compared to the factory headquarters; the endoplasmic reticulum to the warehouse; Golgi apparatus to the distribution center; mitochondria to the power plant; and so on. Try to develop other examples using additional cell organelles. Refer to the *A&P Guide*. The Internet has many cell tutorials. For example, www.cellsalive.com contains an array of interactive links that provide an excellent cell biology resource.

## A. PARTS OF THE CELL

Cells are microscopic units composed of a bubble of fatty material filled with a water-based mixture of molecules and tiny particles. Parts of the cell are called **organelles** (meaning *"small organs"*). Begin your study of the cell by examining a *generalized* cell, such as that in the coloring exercise in **Figure 4-1.** A generalized cell is one with many cell features that are not all found in a single natural cell. Locate each of the listed organelles on a generalized cell model or chart.

☐ 1  The outer boundary of the cell is the **plasma membrane.** It is composed of a double layer (or *bilayer*) of **phospholipid** molecules imbedded with other molecules. Each phospholipid molecule has a polar, hydrophilic ("water loving") *head* made up of phosphate and glycerol, and a nonpolar, hydrophobic ("water fearing") *tail* made up of two fatty acid chains. Imbedded in the membrane are **integral proteins,** which may have additional protein molecules called **peripheral proteins** attached to them on one side of the membrane or the other. **Figure 4-2** shows some of the types of molecules often associated with the plasma membrane. The plasma membrane has many functions, most involving transport and communication between the inside and outside of the cell.

> **Hint** Each membranous organelle described in the following is made of a lipid membrane similar to the cell membrane. Because they are so thin, cell membranes may seem invisible when observed with a light microscope. •

☐ **2** The **nucleus** is a large bubble with a double-walled membrane (**nuclear envelope**) and contains the cell's genetic code.

    ☐ The code is in the form of **deoxyribonucleic acid (DNA)** strands called **chromatin.** Portions of chromatin accept stains readily, giving the nucleus a very dark appearance.

    ☐ A **nucleolus** (literally, "tiny nucleus") is a small area within the nucleus for the synthesis of ribosomal **ribonucleic acid (RNA).** There may be more than one nucleolus (plural *nucleoli*) in a nucleus.

☐ **3** The **endoplasmic reticulum (ER)** is a network of membranous tubes and canals (including the nuclear envelope) winding through the interior of the cell. A *rough ER* is speckled with tiny granules (ribosomes), and a *smooth ER* is not. The ER transports proteins synthesized by ribosomes and other molecules synthesized within its membrane. The ER also manufactures molecules that make up cellular membranes.

## Coloring Exercises: Cell

Use colored pens or pencils to shade in both the figure and the labels. Each red numeral in the figure corresponds to a matching red numeral following the appropriate label.

PLASMA MEMBRANE 1
CENTRIOLE 2
CENTROSOME 3
CHROMATIN 4
CYTOSOL 5
GOLGI APPARATUS 6
LYSOSOME 7
MICROVILLI 8

MITOCHONDRION 9
NUCLEAR MEMBRANE 10
NUCLEOLUS 11
NUCLEAR PORE 12
RIBOSOME 13
ROUGH ER 14
SMOOTH ER 15
VESICLE 16

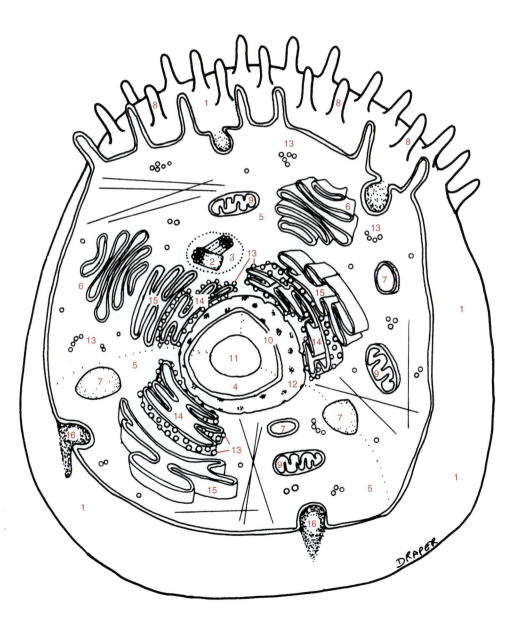

FIGURE 4-1

# *Coloring Exercises:* **Plasma Membrane**

Use colored pens or pencils to shade in both the figure and the labels. Each red numeral in the figure corresponds to a matching red numeral following the appropriate label.

CARBOHYDRATE 1
CARRIER PROTEIN 2
CELL MARKER 3
CHANNEL PROTEIN 4
CHOLESTEROL 5
ENZYME 6
INTEGRAL PROTEIN 7
PERIPHERAL PROTEIN 8
PHOSPHOLIPID,
    NONPOLAR TAIL 9
PHOSPHOLIPID,
    POLAR HEAD 10
RECEPTOR PROTEIN 11

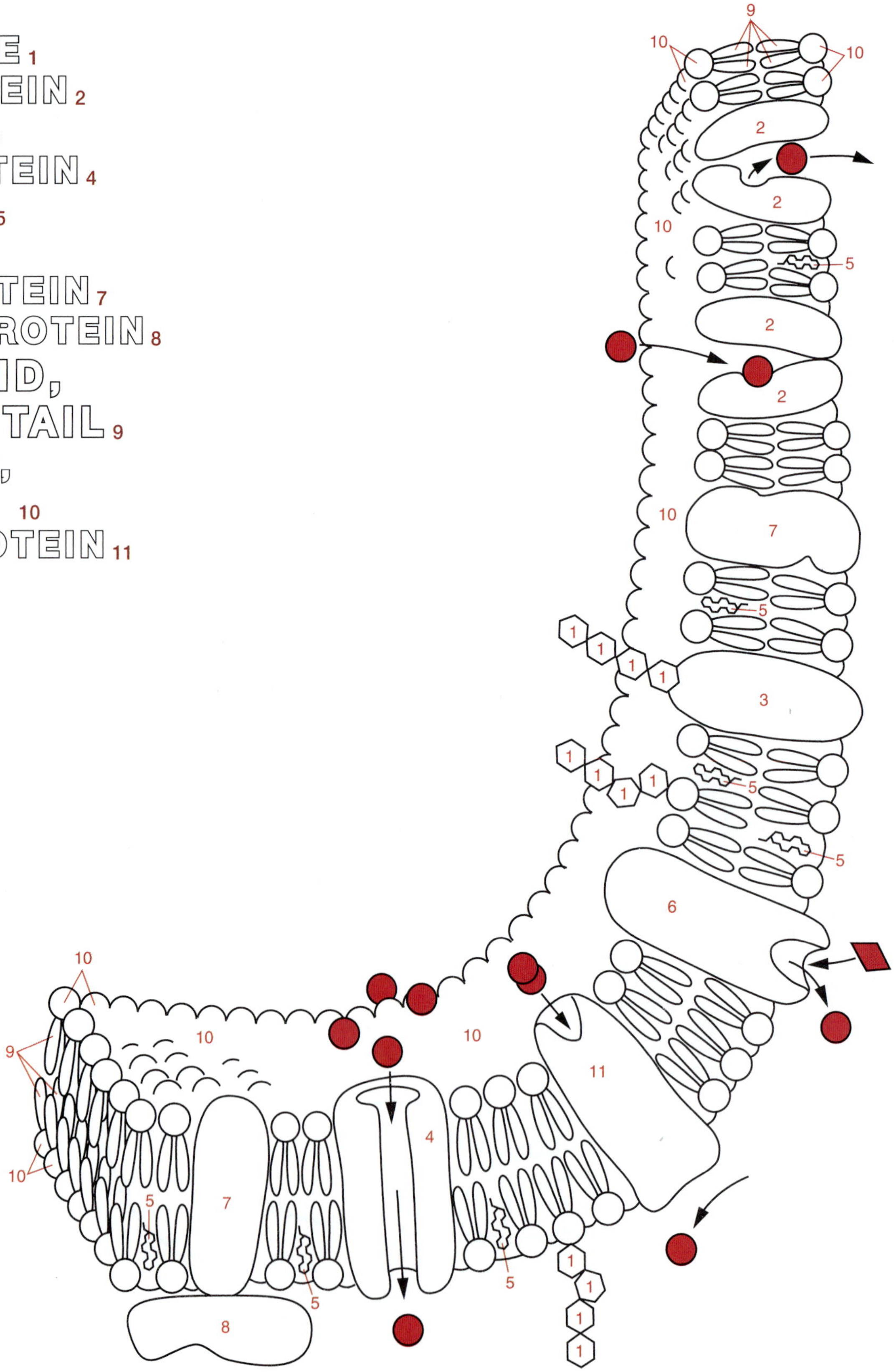

**FIGURE 4-2**

☐ 4 The material within the plasma membrane is the **cytoplasm** (literally, "cell stuff") and includes both the organelles and the liquid, or **cytosol**, surrounding the organelles.

☐ 5 **Ribosomes** are tiny bodies that serve as a site for protein synthesis. Some ribosomes are found on the outer surface of the ER, and some are found scattered elsewhere within the cell.

☐ 6 The **Golgi apparatus,** or **Golgi body,** appears as a stack of flattened sacs. The apparatus receives material from the ER, processes it, and then packages it in tiny vesicles (bubbles) for possible export from the cell.

☐ 7 **Mitochondria** (singular *mitochondrion*) are tiny bodies similar to bacteria that serve as sites for adenosine triphosphate (ATP) synthesis (energy conversion). Mitochondria have an outer membrane, forming a round or oblong capsule, and a folded inner membrane. The folds of the inner membrane are called **cristae.**

☐ 8 **Lysosomes** are vesicles containing digestive enzymes that digest foreign particles and worn or no longer needed cell parts. Also, lysosomes play a role in repairing the plasma membrane.

☐ 9 **Microtubules** are very tiny, hollow beams that form part of the supporting cell skeleton, or *cytoskeleton*. They also form parts of other cell organelles (e.g., flagella, cilia, centrioles, spindle fibers). Other components of the cytoskeleton include **microfilaments** and **intermediate filaments.**

☐ 10 The **centrosome,** or *microtubule organizing center*, is a dense area of cell fluid near the nucleus. The centrosome contains a pair of **centrioles,** cylinders formed by parallel microtubules. A network of microtubules called *spindle fibers* extends from the centrosome during cell division. Spindle fibers distribute DNA equally to the resulting daughter cells.

☐ 11 The cell may have any number of other assorted organelles. **Microvilli** are tiny, fingerlike projections of the cell that increase the membrane's surface area for more efficient absorption. **Cilia** are numerous short, hairlike organelles that propel material along a cell's surface. **Flagella** are single, long, hairlike organelles found in sperm cells to propel them through the female reproductive tract toward the egg. **Vesicles** are membranous bubbles that may be formed by the Golgi apparatus or by the pinching inward of the cell membrane to engulf external substances.

## B. MICROSCOPIC CELL SPECIMEN

Prepare a stained wet-mount specimen of human cheek cells (see Lab Exercise 3—The Microscope for instructions), or obtain another specimen provided by the instructor; try to identify as many cell parts as possible. Use high-power magnification (after first finding the specimen with low-power magnification). Sketch and label your observations in the Lab Report at the end of this exercise.

> ❗ *safety first*
> Be careful when plugging the power cord into the outlet. Always plug it into a receptacle at your station, taking care not to string it over a chair, across an aisle, or in any other dangerous position. Before you turn on the microscope, make sure that the light intensity dial is on the lowest setting. If the light intensity dial is on a high setting, the sudden voltage surge may burn out the lamp.  •

> Hint ▶ Many organelles are very small and can only be seen when properly prepared and examined with a more powerful microscope.  •

## C. INTERPRETING MICROGRAPHS

An **electron microscope** is an instrument that uses a beam of electrons, rather than a beam of light, to form the image of a tiny specimen. *Transmission* electron microscopes send an electron beam through the specimen, similar to the manner in which a light microscope sends a light beam through a specimen. However, the magnifying power and resolution are much greater in the electron microscope. **Resolution** is the ability to distinguish detail. *Scanning* electron microscopes, on the other hand, reflect an electron beam off the specimen. The shadows produced by a scanning electron beam lend a three-dimensional effect.

A **TEM** is a transmission electron micrograph (photograph taken with a transmission electron microscope). **Figure 4-3** shows TEM representations of major cell parts. Use them to help you identify all the cell parts in the labeling exercises in **Figure 4-4,** which shows artists' renderings of TEM images. **Figure 4-5** shows actual TEMs of human cells. This will be a worthwhile challenge to illustrate the difficulty of studying cell structure.

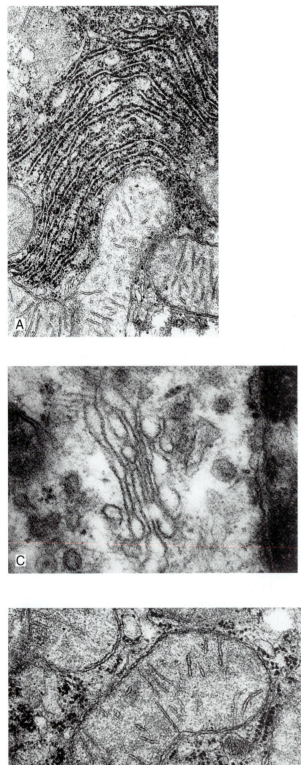

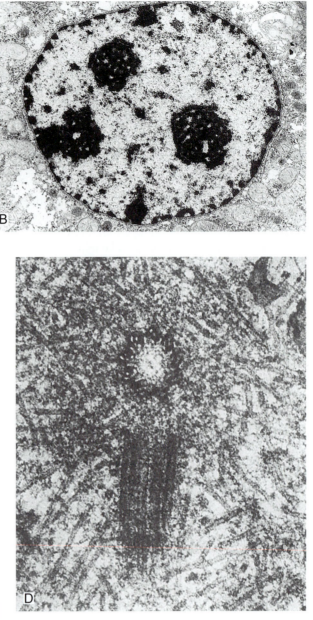

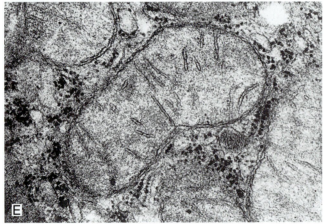

**FIGURE 4-3** Transmission electron micrographs (TEMs) of major organelles. **A,** Rough and smooth endoplasmic reticulum (portions of several mitochondria are also visible). **B,** Nucleus with several nucleoli. **C,** Golgi apparatus. **D,** Centrosome with both centrioles visible. **E,** Mitochondrion with cristae visible.

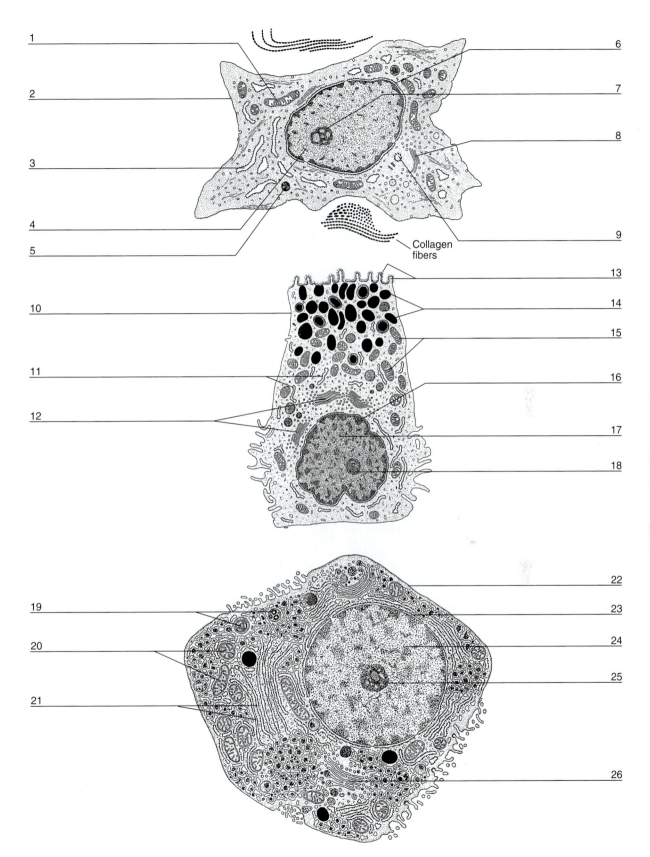

1
2
3
4
5
6
7
8
9

Collagen
fibers

10
11
12
13
14
15
16
17
18

19
20
21
22
23
24
25
26

**FIGURE 4-4** Try to identify the cell organelles labeled in these artist's renderings of TEMs; list them on the lines provided and on the blanks in the Lab Report at the end of this exercise.

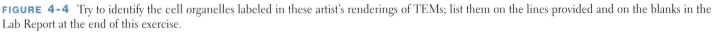

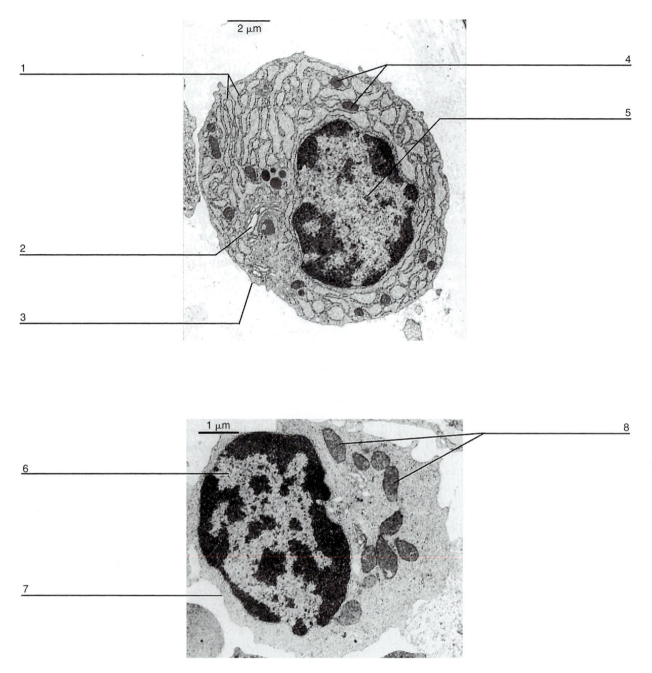

**FIGURE 4-5** Try to identify the cell organelles labeled in these actual TEMs of human cells; list them on the lines provided and on the blanks in the Lab Report at the end of this exercise.

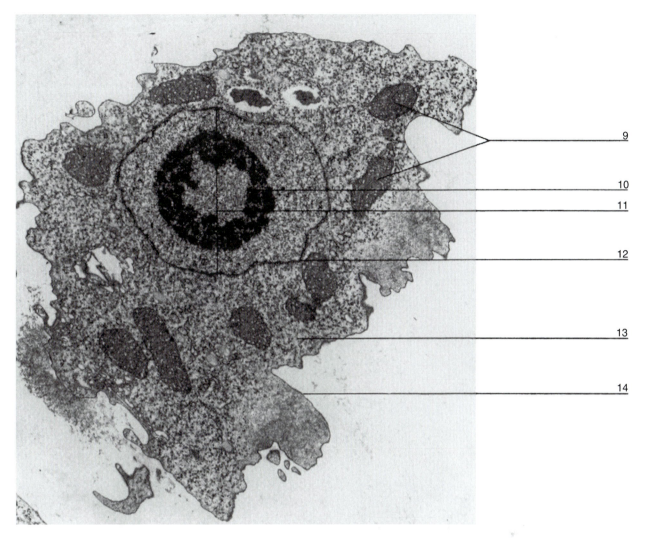

9
10
11
12
13
14

FIGURE 4-5, cont'd

Name: _____ Date: _____ Section: _____

**LAB REPORT 4**
# Cell Anatomy

## Checklist for Cell Model

- ☐ plasma membrane
- ☐ cytosol
- ☐ nucleus
  - ☐ chromatin
  - ☐ nucleolus
- ☐ endoplasmic reticulum
  - ☐ rough ER
  - ☐ smooth ER
- ☐ ribosomes
  - ☐ free
  - ☐ associated with ER
- ☐ Golgi apparatus
  - ☐ secretory vesicles
- ☐ mitochondria
- ☐ lysosomes
- ☐ microtubules
- ☐ centrosome
  - ☐ centrioles
- ☐ microvilli
- ☐ cilia
- ☐ flagella
- ☐ _____
- ☐ _____
- ☐ _____

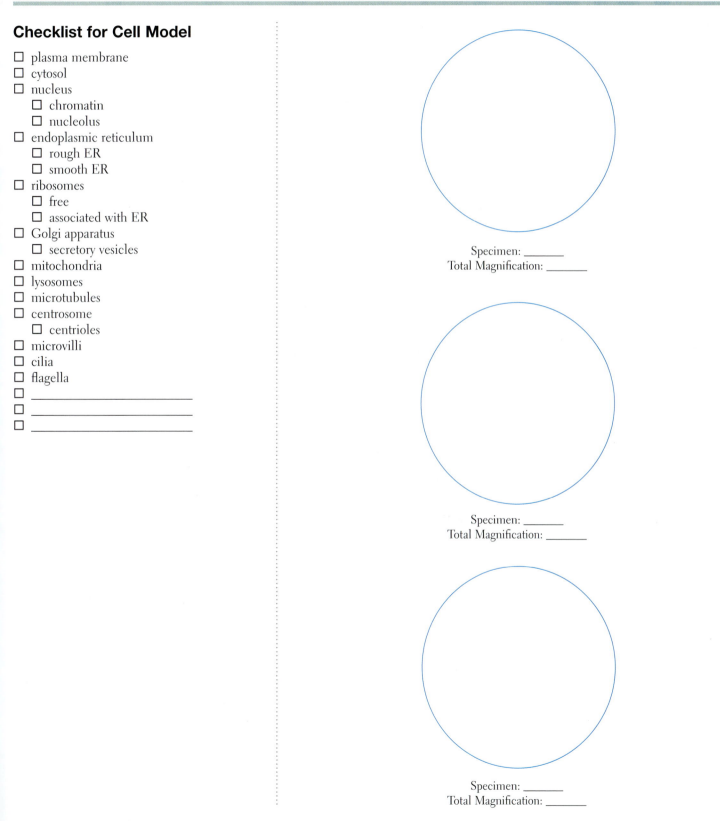

Specimen: _____
Total Magnification: _____

Specimen: _____
Total Magnification: _____

Specimen: _____
Total Magnification: _____

## Figure 4-4 (from p. 39)

1. _____
2. _____
3. _____
4. _____
5. _____
6. _____
7. _____
8. _____
9. _____
10. _____
11. _____
12. _____
13. _____
14. _____
15. _____
16. _____
17. _____
18. _____
19. _____
20. _____
21. _____
22. _____
23. _____
24. _____
25. _____
26. _____

## Figure 4-5 (from pp. 40-41)

1. _____
2. _____
3. _____
4. _____
5. _____
6. _____
7. _____
8. _____
9. _____
10. _____
11. _____
12. _____
13. _____
14. _____

## Matching

1. _____
2. _____
3. _____
4. _____
5. _____
6. _____
7. _____
8. _____
9. _____
10. _____

## Matching (may be used more than once or not at all)

a. plasma membrane
b. centriole
c. endoplasmic reticulum
d. Golgi apparatus
e. lysosome
f. mitochondrion
g. nucleolus
h. nucleus
i. ribosome

1. A double-walled structure containing the cell's genetic code
2. A network of membranous tubes and canals that transports proteins
3. A stack of flattened sacs that process and package proteins
4. Site of manufacture of ribosomal RNA
4. A cylinder formed by parallel microtubules
6. An organelle that serves as the site of protein synthesis
7. A bubble containing digestive enzymes
8. May be *rough* (with ribosomes) or *smooth* (ribosome-free)
9. Allows communication between the internal and external cell environment
10. Forms secretory vesicles

## Fill-in Table (write out the names of the organelles listed, *a* to *i*, in the section above in the appropriate column of the table)

| MEMBRANOUS | NONMEMBRANOUS |
|---|---|
|  |  |
|  |  |
|  |  |
|  |  |
|  |  |
|  |  |

# Transport Through Cell Membranes

Cells use both passive and active processes to transport substances across their membranes. **Passive transport processes** are those that require no metabolic energy from the cell but rely solely on the physical properties of the substances themselves. **Active transport processes** require energy expenditure by the cell to move substances whose physical properties prevent their independent motion. In this exercise, we observe large-scale examples of passive transport processes.

## BEFORE YOU BEGIN

☐ Read the appropriate chapter in your textbook.
☐ Set your learning goals. When you finish this exercise, you should be able to:
  ☐ discuss the nature of diffusion and osmosis
  ☐ use the terms *hyperosmotic*, *hypoosmotic*, and *isosmotic* to compare solutions
  ☐ compare and contrast diffusion, osmosis, and filtration
☐ Prepare your materials:
  ☐ India ink (dropper bottle)
  ☐ compound light microscope
  ☐ microscope slides and coverslips
  ☐ potassium permanganate crystals
  ☐ petri dishes
  ☐ boiling water (100° C) and ice water (0° C)
  ☐ forceps
  ☐ white paper (to place under petri dish)
  ☐ dialysis tubing (8-10 cm, presoaked)
  ☐ string (or tubing clamps)
  ☐ sucrose solutions (15% and 40%)
  ☐ syringe (or pipette) to fill dialysis bags
  ☐ laboratory balance
  ☐ jar or beaker (75-200 mL)
  ☐ distilled water
  ☐ filter paper (circular, large pore)
  ☐ glass funnel
  ☐ flask (to hold funnel and filtrate)
  ☐ copper sulfate (saturated solution)
  ☐ boiled starch solution (10%)
  ☐ charcoal (powdered)
  ☐ Lugol reagent
  ☐ protective gear: gloves, safety eyewear, and apron
☐ Read the directions and safety tips for this exercise *carefully* before starting any procedure.

> **!** *safety first*
> Ask your lab instructor to demonstrate the proper procedure for handling chemicals and glassware and for dealing with spills and other accidents. Wear protective clothing and eyewear. Never point the open end of a test tube toward a person. Be careful to avoid contaminating chemicals: never use a dropper or spatula for more than one type of substance. •

> **!** *safety first*
> Be careful of electrical and heat hazards when dealing with hot water. Review safety procedures with the instructor before beginning the test. •

## A. PHYSICAL BASIS OF PASSIVE TRANSPORT

Naturally occurring **brownian motion** drives passive transport processes. Discovered by Scottish scientist Robert Brown, brownian motion is the constant movement of all particles of matter.

☐ 1 Prepare a wet-mount slide with a drop of India ink.
☐ 2 Observe the slide under high-power magnification. You should be able to see the dye particles moving short, irregular distances. This is brownian motion.

  Because small particles of matter are always bouncing around, brownian motion causes matter to spread (**diffuse**) to areas where there is more room to bounce around. That is, particles tend to move toward an area of lower particle concentration. You can observe this movement by following steps 3 and 4.

☐ 3 Using forceps, place a small crystal of potassium permanganate in the center of a petri dish of calm, boiling water **(Figure 5-1)**. Place a second crystal of the same size in a dish of calm, ice water. Be careful not to disturb the dishes by touching them or causing them to shake.
☐ 4 Observe the two dishes over a period of 30 minutes, carefully noting any changes in the crystals. Answer the questions in the Lab Report at the end of this exercise.

> **!** *safety first*
> Use caution when handling boiling water to avoid injury. •

# *clinical application: osmotic pressure of a solution—cellular effects*

Results similar to those in your osmosis experiment occur when red blood cells come into contact with various solutions that are injected into the blood supply, as in intravenous (IV) therapy. When blood cells are bathed in a solution that is isosmotic to them, they remain unchanged. If a hypoosmotic solution is introduced, the cells experience an inflow of water and usually burst. Bursting of

red blood cells caused by osmosis is called **hemolysis.** If hyperosmotic solution is introduced, the cells lose water and shrivel. Shriveling that results from osmotic loss of water is called **crenation.** Therefore, the concentration of injected materials is critical to the survival of the patient.

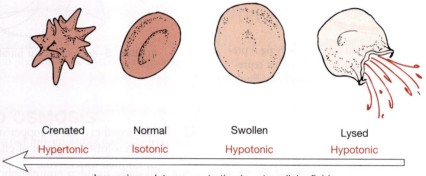

| Crenated | Normal | Swollen | Lysed |
|----------|--------|---------|-------|
| Hypertonic | Isotonic | Hypotonic | Hypotonic |

Increasing solute concentration in extracellular fluid

**FIGURE 5-5**  The effects on red blood cells when changing the osmotic pressure of the extracellular fluid (plasma).

Predict the expected results in these situations:

☐ 1  Randy is a nursing student. He has been asked to intravenously inject his patient with 10 ml of an isosmotic preparation. Mistakenly, he injects his patient with 10 ml of pure water. What is likely to happen to the red blood cells near the site of the injection?

☐ 2  Jennifer is Randy's patient. She is carefully watching him fill an IV bottle with a mixture specially ordered by the physician. She notices that he accidentally fills the bottle with concentrated (10%) salt solution instead of the mixture from the pharmacy. Why should she refuse to allow Randy to attach the IV bottle to her system?

☐ 3  Randy is now assigned to the surgical unit. He is given a piece of living tissue and asked to put it in fluid before taking the sample to the pathology lab. What essential characteristic must such fluid have so that it will not damage any cells in the tissue?

Name: _____ Date: _____ Section: _____

## LAB REPORT 5
# Transport Through Cell Membranes

1. Describe the movement of the India ink particles. Was it uniform or random? Were the particles moving great distances?

| OBSERVATION | ELAPSED TIME | EXTENT OF DIFFUSION IN COLD WATER | EXTENT OF DIFFUSION IN HOT WATER |
|---|---|---|---|
| 1 | 0 minutes | No diffusion of dye particles | No diffusion of dye particles |
| 2 | | | |
| 3 | | | |
| 4 | 30 minutes | | |
| Results | | | |

2. Using potassium permanganate as an example, explain in your own words how brownian motion relates to diffusion.

3. Based on your results in the potassium permanganate experiment, would you state that increased temperature increases or decreases brownian motion?

| BAG | PREDICTED RESULT | INITIAL MASS | FINAL MASS | GAIN OR LOSS OF MASS | EXPLANATION OF RESULTS |
|---|---|---|---|---|---|
| 1 | | | | | |
| 2 | | | | | |
| 3 | | | | | |

4. Did the filtration rate of your mixture become progressively faster or slower, or was it steady? Explain.

| SUBSTANCE | TEST RESULT (PRESENT OR ABSENT) | EXPLANATION OF RESULT |
|---|---|---|
| Charcoal | | |
| Copper sulfate | | |
| Starch | | |

## Matching

1. _____
2. _____
3. _____
4. _____
5. _____
6. _____
7. _____
8. _____
9. _____
10. _____

## Identify

1. _____
2. _____
3. _____
4. _____
5. _____

## Matching (may be used more than once or not at all)

a. brownian motion
b. diffusion
c. filtration
d. hydrostatic pressure
e. hyperosmotic
f. hypoosmotic
g. isosmotic
h. osmosis
i. semipermeable

1. The tendency of matter to spread to areas of lower concentration.
2. Movement through a membrane driven by a hydrostatic pressure gradient.
3. Term that specifically describes the movement of water across a membrane.
4. Term that describes a membrane that allows only some types of particles to pass through it.
5. The natural vibration of particles; it drives diffusion.
6. In the kidney, blood pressure forces some water and solute particles from a blood vessel and into a kidney tubule. What is this type of transport called?
7. A cell is bathed in solution X. The cell quickly shrivels. What term describes solution X (when compared with the cell's fluid)?
8. Solution Q and solution Z have the same amount of impermeant solute particles, but solution Q contains more water. Therefore, solution Q would be _?_ solution when compared with solution Z.
9. A saline (salt) solution is to be injected into a patient. The salt-to-water ratio should be adjusted so that the saline solution is _?_ to the patient's cells.
10. Particles of substance Y move into a cell because there are fewer particles inside the cell than outside. This is an example of _?_.

## Identify (state whether each item is an example of *diffusion, osmosis, filtration,* or *active transport*)

1. Movement of water from an area of lower impermeant solute concentration to an area of higher impermeant solute concentration.
2. Dye particles spread evenly through water.
3. Starch particles pass through a paper membrane.
4. A cell uses energy to "pump" sugar molecules from its external environment.
5. Water in cell moves out into a concentrated salt solution bathing the cell.

# The Cell's Life Cycle

Cells in many parts of the human body divide to produce more cells of the same type. The hereditary information (DNA) contained within the nucleus of a resting *parent* cell must first be **replicated** (copied) and then evenly distributed between the two daughter cells that result from division. The process of distributing genetic material in the parent cell and its subsequent division into offspring cells is termed **mitosis**. Mitosis was named in the late nineteenth century by Walther Flemming, who noticed threadlike structures in cells during cell division (*mitos-* means "threads," and *-osis* means "condition of").

The pinching in of the plasma membrane, and eventual split of the membrane and its contents into two *daughter cells*, is termed **cytokinesis**. Cytokinesis occurs about the same time (or just after) the last phases of mitosis.

*Life cycles* are circular patterns of organisms' life histories. For example, the life cycle of humans would list conception, development, adulthood, and reproduction, and then again list conception and so on for the offspring. Likewise, individual body cells are formed, they reproduce, and their daughter cells continue the cycle of life. In this exercise, we explore the major events of the human cell's life cycle.

## BEFORE YOU BEGIN

☐ Read the appropriate chapter in your textbook.
☐ Set your learning goals. When you finish this exercise, you should be able to:
   ☐ list the major phases of a cell's life cycle
   ☐ describe the principal events of mitosis
   ☐ identify cell parts involved in mitosis
   ☐ explain the importance of mitosis
   ☐ define the term *cytokinesis*
☐ Prepare your materials:
   ☐ chart or model: animal mitosis series
   ☐ colored pencils or pens
   ☐ compound light microscope
   ☐ prepared microslide: whitefish blastula
☐ Read the directions and safety tips for this exercise *carefully* before starting any procedure.

### *Study* TIPS

An effective way to identify the phases of mitosis is to have your lab partner arrange mitosis models or slides randomly and quiz you on the respective phases. During this exercise, concentrate on the appearance of the chromosomes/chromatids. Don't forget to utilize all the resources available to you (textbook, lab manual, prepared mitosis slides, etc.).

## A. INTRODUCTION TO THE CELL LIFE CYCLE

The process of mitosis can be described step by step to make it a little easier to picture. We will divide the whole life cycle into five phases: **interphase, prophase, metaphase, anaphase,** and **telophase**.

Note the major events of each life cycle phase listed in **Figure 6-1**. Try to identify the physical representation of those events in a chart or model of cells at various life cycle stages.

☐ 1 **Interphase** is not a phase of mitosis but is the period between cell divisions. It is not an inactive time, however, because the chromatin replicates during interphase (forming two sister **chromatids** joined at a **centromere**). In anticipation of division, additional cell fluid and organelles are formed during interphase. As **Figure 6-1** shows, an initial growth phase ($G_1$) is followed by the DNA replication or synthesis (S) phase, which is in turn followed by a second growth ($G_2$) phase.

☐ 2 **Prophase** is the first phase of mitosis. During prophase, the nuclear membrane disappears, freeing the chromatin (which first shortens into tiny bodies called **chromosomes**). Also, centriole pairs move to opposite poles of the cell as **spindle fibers** begin to project from them. The spindle fibers extend toward the equator of the cell and may overlap with spindle fibers projecting from the opposite centriole pair. A spindle fiber may also attach to one side of a chromosome's centromere.

> **Hint** The cell life cycle is a continuous process. The cell does not suddenly jump from one phase to another but gradually changes. Therefore, cell models represent a snapshot of each phase at a point where it is most distinct from the phases before and after it. •

☐ 3 **Metaphase** is the period during which the chromosomes (each a pair of replicate chromatids joined at a centromere) line up along the cell's equator (imaginary center plane). Each chromosome now has a pair of spindle fibers attached to it, one from each centriole pair.

☐ 4 **Anaphase** is the phase during which the chromatids split at the centromere, each moving toward an opposite pole along the path of a spindle fiber. At the end of anaphase, each pole of the cell has a full group of 46 single chromosomes. The chromosomes on one side of the cell are replicates of the set of chromosomes on the other side.

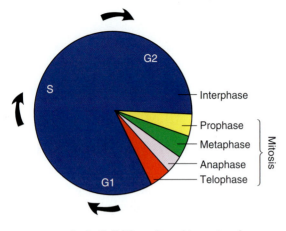

**FIGURE 6-1** Cell life cycle and its major phases.

## B. MICROSCOPIC OBSERVATIONS

In this activity, you will observe cells of the *whitefish blastula* in a prepared slide as in **Figures 6-3** and **6-4**. The blastula is a developmental stage in the growth of many animals. The original cell formed by the joining of the egg and sperm undergoes mitosis many times, at one stage forming a ball of cells called a *blastula*. The cells of the blastula are still undergoing rapid cell division, making it a specimen likely to have many cells at various stages of the cell life cycle.

☐ 1 Obtain a prepared microscope slide (microslide) of whitefish blastula cross-sections (c.s.).

☐ 2 Locate a group of cells by using low-power magnification. There are usually several different cell groups from which to choose on each slide.

☐ 3 Switch to high power and try to locate individual cells in each of the five phases of the cell life cycle. You may need to occasionally switch slides to find all the phases.

☐ 5 **Telophase** is the time during which each side of the cell changes everything to the way it should be during interphase:
  ☐ a nuclear membrane forms around each group of chromosomes
  ☐ the chromosomes uncoil to form long chromatin strands
  ☐ remnants of the spindle fibers disintegrate

☐ 6 During anaphase of mitosis, the separate (but concurrent) process of *cytokinesis* begins. By the end of anaphase, cleavage, or pinching in, of the parent cell is evident. By the end of telophase, complete splitting of the parent cell into two similar daughter cells is complete. Each daughter cell has a nucleus and roughly half of the cytoplasm and organelles of the parent cell.

☐ 7 Review the major events of the cell life cycle by completing the coloring exercise *Cell Life Cycle* (**Figure 6-2**).

The length of time between divisions and the time required for division to take place vary considerably from cell to cell. Even the relative length of different phases varies among individual cells. Cell division can range from 20 minutes to several hours.

> 📌 *landmark characteristics*
> Many prepared whitefish blastula specimens are stained so that the cytoplasm appears pinkish to brown-red (see **Figure 6-3**). The DNA (chromatin, chromosomes) is stained black. Therefore, look for fine, black formations when looking for the major events of mitosis. The plasma membranes and centrioles are so small that they will not appear distinctly in your specimen.
>
> In a single microscopic field you will see many cells, all at different points in the cell life cycle. There is no pattern to the way the cells in different phases are arranged (they are not laid out in order of mitotic phases, for instance). If you are fortunate, you may find at least one example of each phase in a single field. •

> ❗ *safety first*
> Do not forget to check the microscope's power cord for frays and for proper placement. Take care that you don't crack a slide or lens by zooming in with coarse focus without looking from the side. Accidents do not happen when you are careful. •

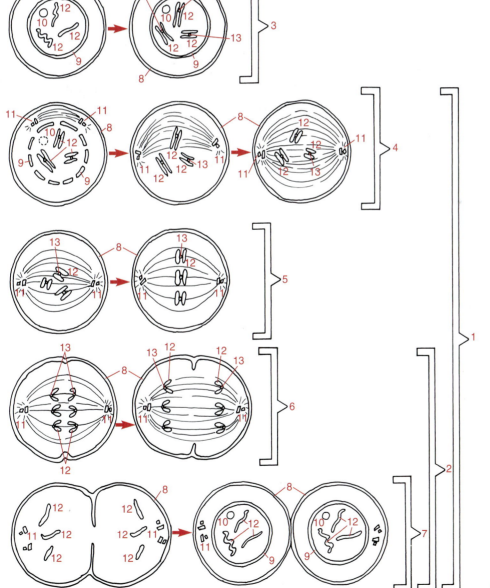

# Coloring Exercises: Cell Life Cycle

Use colored pens or pencils to shade in both the figure and the labels. Each red numeral in the figure corresponds to a matching red numeral following the appropriate label.

MITOSIS 1
CYTOKINESIS 2
INTERPHASE 3
PROPHASE 4
METAPHASE 5
ANAPHASE 6
TELOPHASE 7

CELL MEMBRANE 8
NUCLEAR MEMBRANE 9
NUCLEOLUS 10
CENTRIOLES 11
CHROMATIN/CHROMOSOME 12
CENTROMERE 13

FIGURE 6-2

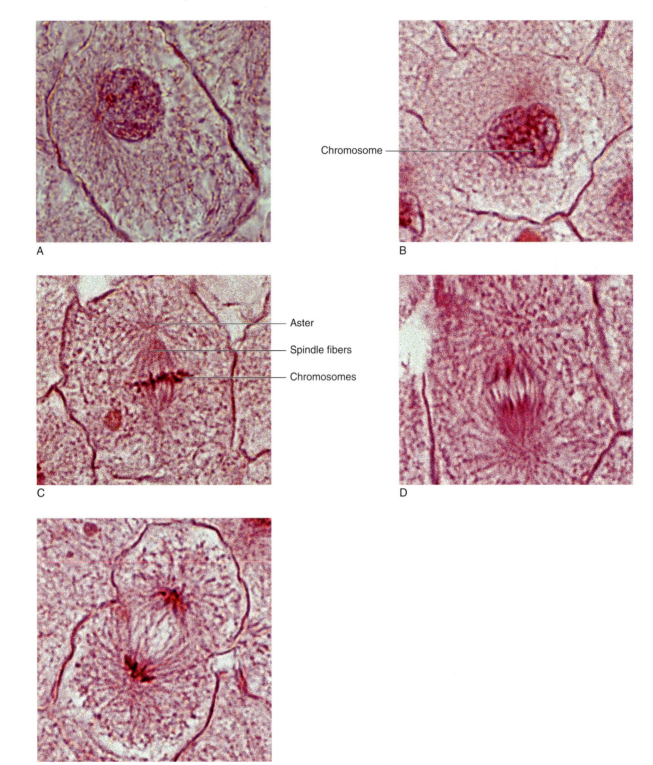

**FIGURE 6-3** The cell life cycle. **A,** Interphase. **B,** Prophase. **C,** Metaphase. **D,** Anaphase. **E,** Telophase.

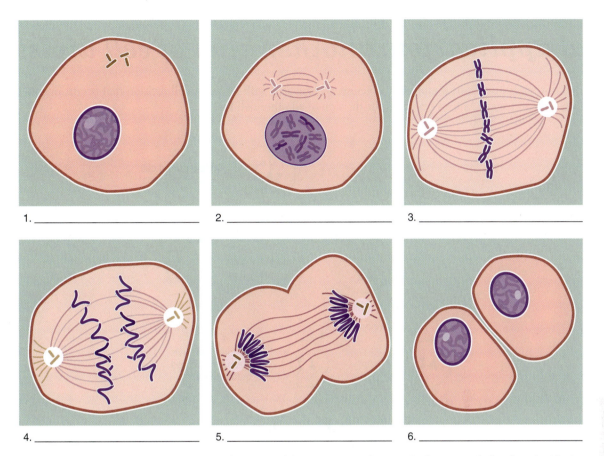

1. _____

2. _____

3. _____

4. _____

5. _____

6. _____

**FIGURE 6-4** Examples of whitefish cells. Label each with the name of the appropriate phase on the lines provided and on the blanks in the Lab Report at the end of this exercise.

# *clinical application: another kind of cell division*

Meiosis is a process distinct from mitosis (Figure 6-5). Meiosis is associated with a type of cell division that occurs during the formation of reproductive cells (*sperm* and *eggs*). Mitosis occurs in the division of all other cell types.

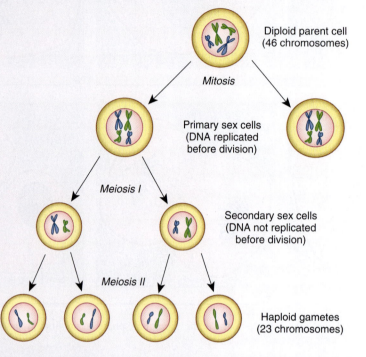

**FIGURE 6-5** Meiosis. Meiotic cell division takes place in two steps: meiosis I and meiosis II. Meiosis is called *reduction division* because the number of chromosomes is reduced by half (from the diploid number to the haploid number).

Meiosis results in daughter cells that have only half the number of chromosomes that other cells, including the parent cells, have. That is, during meiosis, a parent cell with 46 chromosomes (the diploid number) produces daughter cells that have 23 chromosomes each (the haploid number). This must occur so that when the sperm and egg unite during conception, the newly formed cell has 46 chromosomes (23 from the sperm plus 23 from the egg). Thus, the offspring has equal amounts of hereditary information from each parent.

Theoretically, what would happen if meiosis did not occur and sperm and egg cells could only form using mitosis?

What would happen if *all* your body cells divided using meiosis instead of mitosis?

Name: _____ Date: _____ Section: _____

**LAB REPORT 6**
# The Cell's Life Cycle

Sketch your observations of particularly clear examples of mitotic phases in whitefish cells.

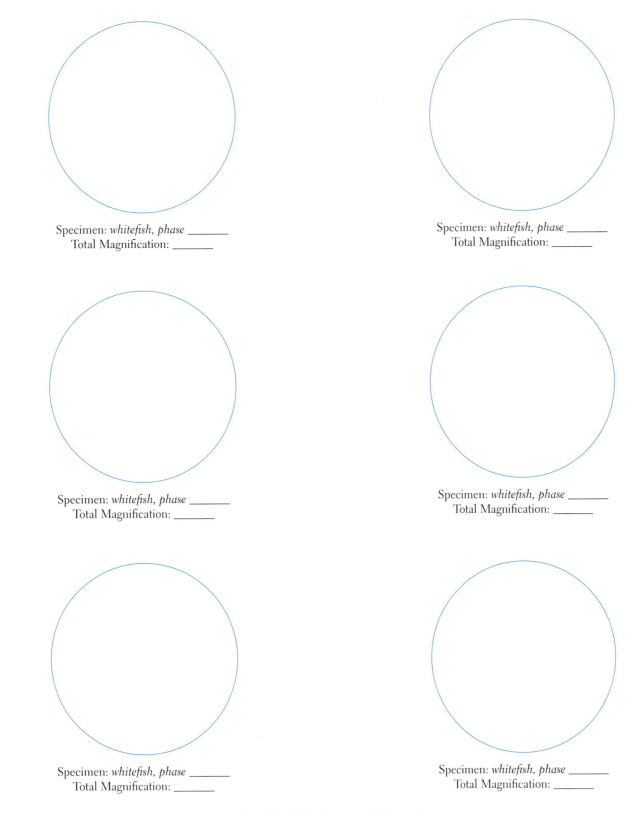

Specimen: *whitefish, phase* _____
Total Magnification: _____

Specimen: *whitefish, phase* _____
Total Magnification: _____

Specimen: *whitefish, phase* _____
Total Magnification: _____

Specimen: *whitefish, phase* _____
Total Magnification: _____

Specimen: *whitefish, phase* _____
Total Magnification: _____

Specimen: *whitefish, phase* _____
Total Magnification: _____

**TABLE 7-1** **Examples of Major Membranous Epithelial Tissues**

| TISSUE | LOCATION | FUNCTION |
|---|---|---|
| Simple squamous | Alveoli of lungs<br>Lining of blood and lymphatic vessels (called *endothelium;* classed as connective tissue by some histologists)<br>Surface layer of pleura, pericardium, peritoneum (called *mesothelium;* classed as connective tissue by some histologists) | Absorption by diffusion of respiratory gases between alveolar air and blood<br>Absorption by diffusion, filtration, osmosis<br>Absorption by diffusion and osmosis; also secretion |
| Stratified squamous<br>• Nonkeratinized<br>• Keratinized | Surface of mucous membrane lining mouth, esophagus, and vagina<br>Surface of skin (epidermis) | Protection<br>Protection |
| Simple cuboidal | In many types of glands and their ducts; also found in ducts and tubules of other organs, such as the kidney | Secretion; absorption |
| Pseudostratified columnar | Surface of mucous membrane lining trachea, large bronchi, nasal mucosa, and parts of male reproductive tract (epididymis and vas deferens); lines large ducts of some glands (e.g., parotid) | Protection |
| Simple columnar | Surface layer of mucous lining of stomach, intestines, and part of respiratory tract | Protection; secretion; absorption; moving of mucus (by ciliated columnar epithelium) |
| Transitional | Surface of mucous membrane lining urinary bladder and ureters | Permits stretching |

> **⚠ safety first**
>
> Avoid the hazards associated with frayed power cords and broken glass slides. Do not reach for anything while looking into the ocular lenses, or you may knock over something (or someone). Remember to make sure the intensity dial is on its lowest setting before plugging in the microscope. •

For each of the epithelial tissues listed, examine at least one example in a prepared slide. Practice looking for the key characteristics by which each type can be classified. Sketch some representative examples in the Lab Report at the end of this exercise.

>  In many (not all) prepared epithelia, the cytoplasm will appear either as a clear/cloudy area or a very pale pink. The cell membranes are usually faint but may be visible as medium-pink lines. The nuclei are often stained dark pink to violet or bluish. •

☐ 1 **Simple squamous epithelium**—As the name implies, simple squamous epithelium is a single layer of flattened cells **(Figure 7-4)**. This type of epithelium forms the very thin lining found in the blood vessels, in the alveoli (air sacs) of the lungs, and in other areas where thin membranes are required. Because it is so thin, simple squamous is well adapted for diffusion or filtration of water, gases, and other substances.

> **📌 landmark characteristics**
>
> Viewed from the side, as in a cross-section, simple squamous epithelium looks like a thin line of cells—often with distinguishable nuclei (see **Figure 7-4, A**). Viewed as a sheet from above, simple squamous epithelium looks like a two-dimensional layout of polygonal or rounded "tiles," each with a central nucleus (see **Figure 7-4, B**). •

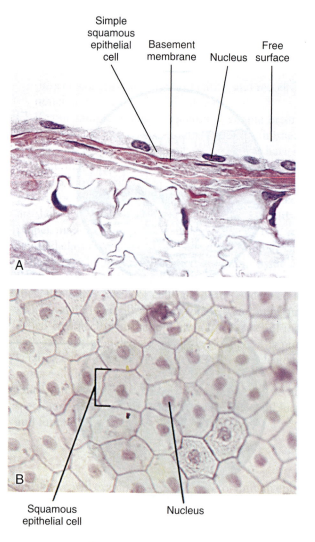

**FIGURE 7-4** Simple squamous epithelium.

2 **Stratified squamous epithelium**—This tissue type is composed of multiple layers of cells: columnar cells along the basement membrane, topped by cuboidal cells, then by squamous cells. Cells divide in the columnar layer and are pushed upward, where they are distorted into cuboidal, then squamous, cells. The cells on the surface slough off but are continually replaced by cells moving up from the bottom layer. Because of its thickness and its constant renewal, it is well adapted for protection. For example, stratified squamous epithelium is found in the outer part of the skin and the mucous linings of the mouth, vagina, and esophagus.

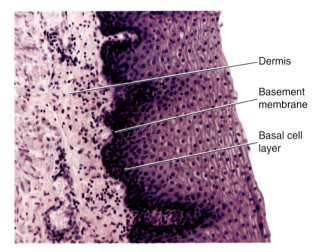

FIGURE 7-5  Stratified squamous (nonkeratinized) epithelium.

### 📌 *landmark characteristics*

- Nonkeratinized stratified squamous epithelium (Figure 7-5) typically has a dense concentration of nuclei (of columnar and cuboidal cells) near the basement membrane, becoming less dense toward the free surface. Most of the squamous cells near the free surface should have identifiable nuclei.
- Keratinized stratified squamous samples will be from the epidermis (outer layer) of the skin. They will look similar to nonkeratinized specimens, except that they have distinct additional layers overlying the top layers of squamous cells (Figure 7-6). This layer has no distinguishable nuclei. •

FIGURE 7-6  Stratified squamous (keratinized) epithelium.

□ 3 **Simple cuboidal epithelium**—Composed of a single layer of almost cubic cells, this type is found in secreting organs such as glands. Simple cuboidal epithelium also forms the kidney tubules, where it is specialized for water reabsorption and ion movement (Figure 7-7).

### 📌 *landmark characteristics*

The cuboid shape of cells is easily seen in cross-sections of kidney tubules (see Figure 7-7). Because the sample is formed by many tubules cut at an angle, you will see many circles and loops made of simple cuboidal epithelium. Similar specimens are seen in thyroid tissue or other glands and glandular ducts. •

cup has no wine but rather **mucus**, which the goblet cells produce in great quantity. Mucus has many functions, including the lubrication and protection of the epithelial lining.

□ 4 **Simple columnar epithelium**—Forming linings specialized for absorption and secretion, simple columnar epithelium is found in many parts of the body (Figure 7-8). For example, this type of epithelium lines portions of the reproductive tract, digestive tract, excretory ducts, and respiratory tract. A special cell that is often interspersed among the other columnar cells is the **goblet cell**. The goblet cell resembles its namesake, the wine goblet, in that it has a large, cuplike vesicle that may open onto the free surface. The

### 📌 *landmark characteristics*

Simple columnar sheets often line cavities with deeply folded or grooved walls. The specimen, then, will appear to zigzag when viewed in a cross-section. One surface of the sheet is always free, however, even though the free surfaces may fold back and touch one another. A goblet cell is easily recognizable by the very large bubble (vesicle) in the center or near the top (see Figure 7-8). Because the vesicle contains clear, unstained mucus, it will appear more lightly colored than the surrounding material. •

### landmark characteristics

The collagen fibers of hyaline cartilage are not distinct in the matrix, which has a smooth, pinkish, or lavender appearance in many preparations. The chondrocytes are usually pink to violet and appear to have shriveled within their respective lacunae. Because of chondrocyte shrinkage, a clear ring appears around the inside of many lacunae (see **Figure 8-9**).  •

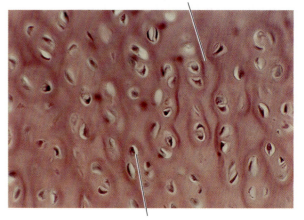

Collagen fiber matrix

Cells in lacuna

**FIGURE 8-10** Fibrocartilage.

☐ **6 Fibrocartilage**—The name of this tissue indicates its high concentration of collagen fibers. These fibers give the tissue a distinctive fibrous appearance **(Figure 8-10)**. Fibrocartilage has a more rigid, less rubbery consistency than other cartilage types. Fibrocartilage forms the disks between vertebrae and may be found at other semimovable joints.

### landmark characteristics

Fibrocartilage may appear alongside hyaline cartilage and sometimes looks very much like it. However, the distinct fibrous appearance of fibrocartilage's matrix is the determining factor (see **Figure 8-10**).  •

☐ **7 Elastic cartilage**—As its name implies, elastic cartilage has a large proportion of elastin fibers in its matrix **(Figure 8-11)**. Elastic cartilage is found in structures in which springiness is desirable in the support material. For example, the elasticity of the pinna (ear flap) is provided by elastic cartilage.

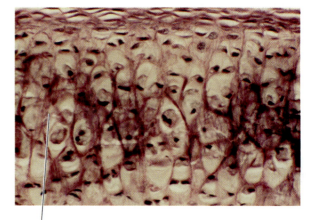

Elastin fibers

**FIGURE 8-11** Elastic cartilage.

### landmark characteristics

In many preparations, elastin fibers stain very darkly. This makes their presence easy to detect and allows one to distinguish elastic cartilage without any problem (see **Figure 8-11**).  •

☐ **8 Compact bone**—Compact bone is formed by solid, cylindrical units called **osteons** packed tightly together. The osteon, or *haversian system*, consists of multiple concentric layers of hard bone matrix, with cells sandwiched between each layer **(Figure 8-12)**. Each layer is a **lamella** (plural, *lamellae*). **Osteocytes** are literally trapped within lacunae between the lamellae. The osteocytes were once active **osteoblasts** but have trapped themselves in the solid matrix they formed. The lamellae are centered around the **central** (haversian) **canal's** blood vessels. The cells transport materials to and from the canal by way of tiny **canaliculi** ("small canals") that connect the osteocytes to one another and to the canal.

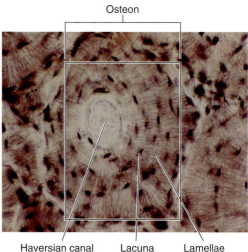

Osteon

Haversian canal      Lacuna      Lamellae

**FIGURE 8-12** Compact bone.

### 🎯 *landmark characteristics*

A cross-section of compact bone has rings of lamellae surrounding several adjacent central canals. The lamellae resemble rings in an onion slice. The central canals are either clear or nearly black, the lamellae are buff to orange, and the osteocytes are brown or black. The canaliculi often appear as wavy hairlines radiating from the lacunae (see **Figure 8-12**). •

☐ 9 **Cancellous bone and hematopoietic tissue**—Cancellous bone is easily identified by its open, lattice-like structure. Thin plates of bone matrix, with a scattering of osteocytes trapped within lacunae, form structural beams that have great strength despite the open spaces. These beams of hard bone are called **trabeculae**. Because cancellous bone has open spaces, it is sometimes called *spongy bone*. This name can be misleading because one might think spongy bone is as soft as a bath sponge. It is not soft at all because it has hard trabeculae. The spaces are filled with *hematopoietic* or *myeloid tissue*, a special type of blood tissue that produces new blood cells **(Figure 8-13)**. Hematopoietic tissue is also called *red bone marrow*.

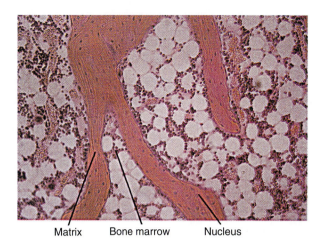

Matrix    Bone marrow    Nucleus

**FIGURE 8-13** Cancellous bone and hematopoietic tissue.

### 🎯 *landmark characteristics*

Cancellous bone is distinguished by its rather disorganized array of trabecular beams of bone surrounded by myeloid tissue. The bone pieces may look like slivers of compact bone, with lamellae that often do not form complete circles. The myeloid tissue is a scattering of blood cells, which appear as tiny, dark circles. Myeloid (hematopoietic) tissue may also have a netlike formation of very thin collagen fibers called *reticular fibers*. In some preparations, the bone tissue is pink and the myeloid cells are dark red (see **Figure 8-13**). •

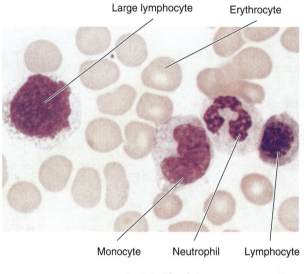

Large lymphocyte    Erythrocyte

Monocyte    Neutrophil    Lymphocyte

**FIGURE 8-14** Blood tissue.

☐ 10 **Blood**—Blood tissue is a fluid matrix connective tissue characterized by a variety of cell types **(Figure 8-14)**. Tiny **red blood cells (RBCs)**, tinier fragments called **platelets**, and larger **white blood cells (WBCs)** may all be seen in a blood smear preparation. Of course, blood is found within the circulation vessels and has many functions. Blood transports and exchanges materials, serves in immune protection of the body, and helps regulate body temperature, among other functions. We will study blood in more detail in Lab Exercise 34—Blood.

## C. GROSS TISSUE SAMPLES

Gross (large) examples of connective tissues often give one a more complete mental image of connective tissue types. If gross specimens are available, explore the texture, consistency, and other physical characteristics of the connective tissue listed.

### 🎯 *landmark characteristics*

A prepared blood smear is a drop of blood smeared on a slide and stained. The RBCs are the more numerous, smaller cells stained pink to orange-red. RBCs have no nuclei. The WBCs are much larger, with distinct, often distorted nuclei. Lab Exercise 34—Blood has more complete details (see **Figure 8-14**). •

### ⚠️ *safety first*

Any animal tissue at room temperature can harbor growing colonies of microorganisms. As a safety precaution, treat the samples as if they are contaminated with harmful microbes. Use gloves or dissection tools when handling the samples and disinfect any surfaces that they touch. •

Fill in Table 8-2 with the physical characteristics of different types of tissue (Figure 8-15).

TABLE 8-2    Tissue Characteristics

| TYPE | PHYSICAL CHARACTERISTICS |
| --- | --- |
| Dense regular fibrous | |
| Dense irregular fibrous | |
| Loose fibrous (areolar) | |
| Adipose | |
| Hyaline cartilage | |
| Fibrocartilage | |
| Elastic cartilage | |
| Compact bone | |
| Cancellous bone | |
| Blood | |

# Coloring Exercises: Connective Tissues

Use colored pens or pencils to shade in both the figure and the labels. Each red numeral in the figure corresponds to a matching red numeral following the appropriate label.

LOOSE, ORDINARY
(AREOLAR) 1
RETICULAR 3
HYALINE CARTILAGE 5
ELASTIC CARTILAGE 7

ADIPOSE 2
DENSE FIBROUS
(REGULAR) 4
FIBROCARTILAGE 6
COMPACT BONE 8

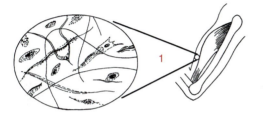
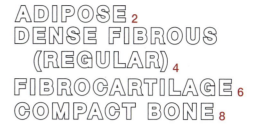
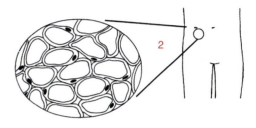
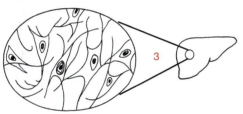

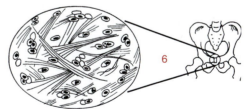
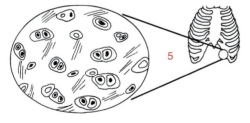
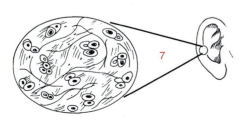
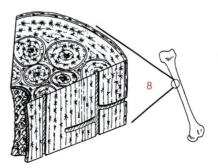

**FIGURE 8-15** Types of connective tissues.

Name: _____ Date: _____ Section: _____

Name: _____ Date: _____ Section: _____

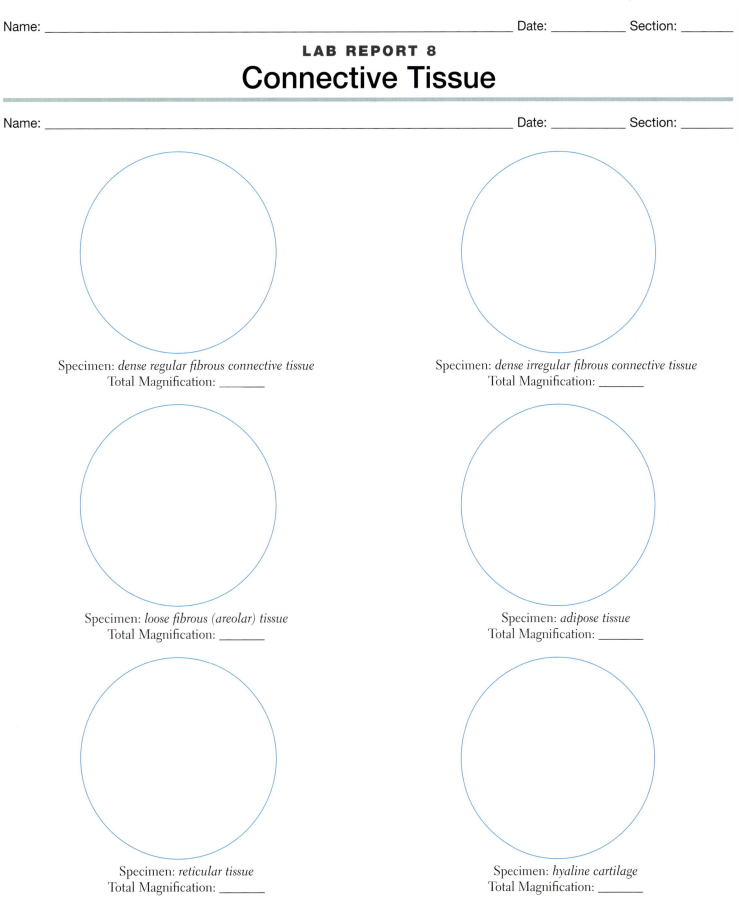

Specimen: *dense regular fibrous connective tissue*
Total Magnification: _____

Specimen: *dense irregular fibrous connective tissue*
Total Magnification: _____

Specimen: *loose fibrous (areolar) tissue*
Total Magnification: _____

Specimen: *adipose tissue*
Total Magnification: _____

Specimen: *reticular tissue*
Total Magnification: _____

Specimen: *hyaline cartilage*
Total Magnification: _____

Specimen: *fibrocartilage*
Total Magnification: _____

Specimen: *elastic cartilage*
Total Magnification: _____

Specimen: *compact bone*
Total Magnification: _____

Specimen: *cancellous bone*
Total Magnification: _____

Specimen: *blood smear*
Total Magnification: _____

Name: _____ Date: _____ Section: _____

## Figure 8-2 (from p. 73)

1. _____
2. _____
3. _____
4. _____
5. _____

## Figure 8-3 (from p. 74)

1. _____
2. _____
3. _____
4. _____
5. _____
6. _____

## Matching

1. _____
2. _____
3. _____
4. _____
5. _____
6. _____
7. _____
8. _____
9. _____
10. _____

## Matching (may be used more than once)

a. dense fibrous connective tissue
b. adipose
c. hyaline cartilage
d. fibrocartilage
e. elastic cartilage
f. compact bone
g. cancellous bone

1. Cartilage type with a great deal of collagen in the matrix
2. Tissue type that is actually modified areolar tissue
3. Tissue type that stores lipid molecules
4. Tissue type composed of haversian systems
5. The most common type of cartilage
6. Strong tissue that forms tendons and ligaments
7. Tissue type associated with red bone marrow
8. Tissue type found in the external ear
9. Connective tissue that forms the disks between vertebrae
10. Tissue type that forms hard mineral trabeculae

**TABLE 9-1    Examples of Major Muscle and Nerve Tissues**

| TISSUE | LOCATION | FUNCTION |
|---|---|---|
| **Muscle** | | |
| Skeletal (striated voluntary) | Muscles that attach to bones<br>Extrinsic eyeball muscles<br>Upper third of esophagus | Movement of bones<br>Eye movements<br>First part of swallowing |
| Smooth (nonstriated, invol-untary, or visceral) | In walls of tubular viscera of digestive, respiratory, and genitourinary tracts<br>In walls of blood vessels and large lymphatic vessels<br><br>In ducts of glands<br>Intrinsic eye muscles (iris and ciliary body)<br>Arrector muscles of hairs | Movement of substances along respective tracts<br>Change diameter of blood vessels, thereby aiding in regulation of blood pressure<br>Movement of substances along ducts<br>Change diameter of pupils and shape of lens<br>Erection of hairs (goose pimples) |
| Cardiac (striated involuntary) | Wall of heart | Contraction of heart |
| **Nervous** | Brain<br>Spinal cord<br>Nerves | Excitability<br>Conduction |

☐ **2 Cardiac muscle** is also known as *striated involuntary muscle*. This tissue also has striations, though less distinct than in skeletal muscle **(Figure 9-2)**. Cardiac muscle is involuntary in the sense that its contraction is not under the control of the voluntary motor nerves. Cardiac muscle is found only in the walls of the heart. Because it must encircle and compress the heart chambers with great strength to pump blood, cardiac muscle requires some features not found in other muscle types. For example, the individual fibers are branched, allowing the fibers to mesh with other cells at different layers. Also, cardiac muscle fibers are fused end to end by **intercalated disks**. This branching and fusing gives a group of cells the ability to functionally imitate a giant cell encircling one or more chambers of the heart.

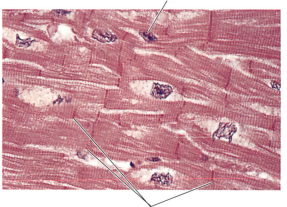

Nucleus

Intercalated disks

**FIGURE 9-2** Cardiac muscle.

> 📌 **landmark characteristics**
> Cardiac muscle striations are less distinct than those in skeletal muscle. Cardiac fibers have single nuclei, usually have branches, and do not taper at their ends. Many fibers attach end to end via intercalated disks, which appear as fine, dark (sometimes purple) lines at a right angle to a seemingly continuous fiber (see **Figure 9-2**). •

☐ **3 Smooth muscle** gets its name because it has no distinct striations. Like cardiac muscle, it also is an involuntary muscle type. Smooth muscle is found in the walls of hollow organs, such as digestive organs and blood vessels. This tissue type is composed of long, threadlike cells, each with a single nucleus. The cells are generally parallel with one another and with the edge of the wall in which they are embedded **(Figure 9-3)**.

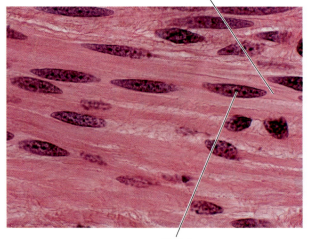

Smooth muscle cell

Nuclei

**FIGURE 9-3** Smooth muscle.

### landmark characteristics

Unlike other muscle cells, smooth muscle fibers are unstriated and have single nuclei. In some preparations the cells are pink and the nuclei purple or black. At first glance, you may confuse smooth muscle tissue with dense fibrous (regular) connective tissue. Smooth muscle is not wavy, as dense fibrous tissue sometimes is, and has a more even distribution of nuclei than dense fibrous tissue (see **Figure 9-3**).  •

## B. NERVE TISSUE

Nerve tissue composes organs of the nervous system: the brain, spinal cord, and nerves. Two basic types of cells are found in this tissue: **neurons** (impulse-conducting cells) and **glia** (support cells). Neurons are large cells with nucleated *bodies* and projections called *axons* and *dendrites* **(Figure 9-4)**. There are many types of glia, or *neuroglia*, as they are sometimes called, but they are generally smaller than neurons, which they outnumber by several times. Glia surround and support neurons physically or biochemically.

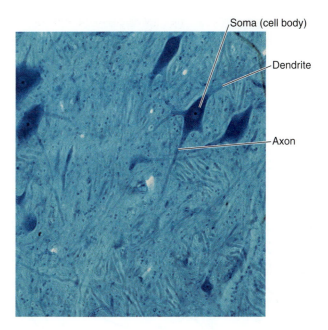

**FIGURE 9-4** Nerve tissue.

### landmark characteristics

In a smear of spinal cord tissue, the neurons and glia are scattered randomly on the slide. The neurons are extremely large, each with a body having a single nucleus. Neuron projections crisscross throughout. Glia appear as tiny, dark dots (see **Figure 9-4**).  •

### *Study* TIPS

Refer to the Landmark Characteristics boxes and familiarize yourself with the unique characteristics that define each type of muscle tissue. The use of flash cards or review cards is an excellent strategy to learn the various types of muscle and nerve tissue. Obtain photos or illustrations of the different types of muscle and nerve tissue. Place the photo or illustration on one side of an index card. On the opposite side of the card, put the name of the tissue. You can also add additional information such as unique characteristics or location in the body.

## Coloring Exercises: Muscle and Nerve Tissues

Use colored pens or pencils to shade in both the figure and the labels. Each red numeral in the figure corresponds to a matching red numeral following the appropriate label.

SKELETAL MUSCLE₁
SMOOTH MUSCLE₂

CARDIAC MUSCLE₃
NERVE TISSUE₄

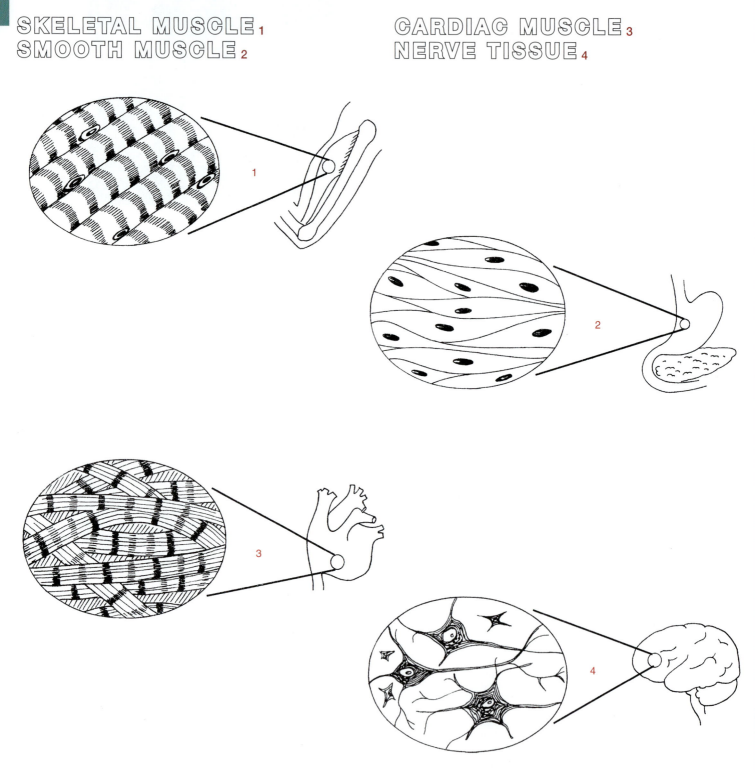

**FIGURE 9-5**  Muscle and nerve tissue drawings.

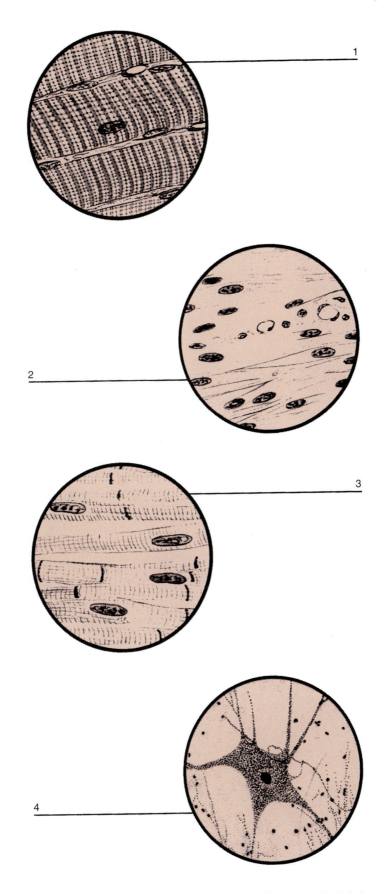

1

2

3

4

**FIGURE 9-6** Try to identify these tissue types and label them on the lines provided and on the Lab Report at the end of this exercise.

Name: _____ Date: _____ Section: _____

## LAB REPORT 9
# Muscle and Nerve Tissue

Specimen: *skeletal muscle*
Total Magnification: _____

Specimen: *cardiac muscle*
Total Magnification: _____

Specimen: *smooth muscle*
Total Magnification: _____

Specimen: *spinal cord smear*
Total Magnification: _____

**Figure 9-6** (from p. 91)

1. _____

2. _____

3. _____

4. _____

**Multiple Choice**

1. _____

2. _____

3. _____

4. _____

5. _____

6. _____

7. _____

8. _____

**Multiple Choice** (choose the one best response)

1. A voluntary type of muscle tissue is
   a. skeletal
   b. cardiac
   c. smooth
   d. a and b are correct
   e. a and c are correct
2. An involuntary muscle tissue is
   a. skeletal
   b. cardiac
   c. smooth
   d. b and c are correct
   e. a and c are correct
3. Which of these muscle tissues is striated?
   a. skeletal
   b. cardiac
   c. smooth
   d. a and b are correct
   e. all are correct
4. Which of these is most likely to be found in the wall of the urinary tract?
   a. skeletal
   b. cardiac
   c. smooth
5. Cardiac muscle fibers are connected end to end by means of
   a. Velcro
   b. intercalated disks
   c. striations
   d. branching
6. Which of these cells is likely to have multiple nuclei?
   a. skeletal muscle
   b. cardiac muscle
   c. smooth muscle
   d. neuron
   e. a and c are correct
   f. a, c, and d are correct
7. Nerve tissue contains cells called
   a. fibers
   b. neurons
   c. glia
   d. striated cells
   e. b and c are correct
8. Nerve tissue forms the bulk of the
   a. brain
   b. heart
   c. digestive tract
   d. vertebrae
   e. a and d are correct

# The Skin

The **skin** is the primary organ of the integumentary system and is the largest organ of the body. Forming the outer protective covering of the body, the skin is a continuous sheet of **cutaneous membrane**.

## BEFORE YOU BEGIN

☐ Read the appropriate chapter in your textbook.
☐ Set your learning goals. When you finish this exercise, you should be able to:
  ☐ describe the major structures of the skin and identify their functions
  ☐ identify important skin structures in a diagram, model, and prepared specimen
  ☐ compare and contrast features of thick skin and thin skin
  ☐ describe the gross structure of skin in a preserved mammalian specimen
☐ Prepare your materials:
  ☐ colored pencils or pens
  ☐ model or chart of a skin cross-section
  ☐ compound light microscope
  ☐ prepared microslides:
    ☐ thin skin section
    ☐ thick skin section
    ☐ hair follicles c.s.
  ☐ preserved specimen: cat or fetal pig (unskinned)
  ☐ dissection tools and trays
  ☐ protective gear: gloves, safety eyewear, and apron (if specimen is available)
  ☐ storage container (if specimen is to be reused)
☐ Read the directions and safety tips for this exercise *carefully* before starting any procedure.

### *Study* TIPS

Review the major tissues types of the body. The skin is an organ, and all four major tissues types are represented. Refer to Lab Exercises 7 to 9 and construct a table or chart that defines the role of each tissue type within the skin. To understand the relationship of the accessory organs (hair, nails, and glands), you may want to develop a concept map that illustrates the relationship of these structures and the function of the skin.

> ! *safety first*
> Don't forget the rules for safe use of the microscope. •

## A. BASIC SKIN STRUCTURE

First in a model and then in prepared microscopic specimens, identify the elements of skin structure described in the following steps:

☐ 1 The skin has two distinct layers. The superficial layer is a sheet of keratinized stratified squamous epithelium called the **epidermis**. The epidermis is itself divided into distinct histological regions or **strata** (meaning "layers"):
  ☐ **Stratum basale** is the deepest stratum of the epidermis. It consists of a single sheet of columnar cells that continue to divide. As the daughter cells are formed, they are pushed upward, becoming part of the next stratum.
  ☐ **Stratum spinosum** is noted for its multilayer of distorted ("spined") cells. The cells become distorted as they are pushed up from the deeper stratum basale. Stratum basale and stratum spinosum together are often called stratum germinativum.
  ☐ **Stratum granulosum** is superficial to stratum spinosum. It contains flattened cells pushed up from the deeper strata. As the cells are pushed up through this stratum, they form the protein granules that give it the name granulosum. By the time the cells leave this stratum, they have died.
  ☐ **Stratum lucidum** (meaning "light layer") is a very thin layer present only in **thick skin**. Thick skin is found only in high-wear areas such as the palms and soles. The more flexible **thin skin** is found over most other areas of the body. This stratum's name comes about because it is translucent, allowing light to pass through it easily.
  ☐ **Stratum corneum** is the layer of dead, keratinized tissue already identified in Lab Exercise 7—Epithelial Tissue. Stratum corneum is extremely thick in thick skin, providing a great deal of protection. Stratum corneum protects deeper tissues from mechanical injury, from inward or outward diffusion of water and other molecules, and from invasion by microorganisms.

☐ 2 The layer of skin deep to the epidermis is a sheet of irregular fibrous connective tissue called the **dermis**. The dermis is usually much thicker than the epidermis. Like most connective tissues, the dermis has a scattering of blood vessels and nerves. The blood vessels supply both the dermis and epidermis. Because blood cools when it travels through the skin, the body varies the amount of blood sent to the skin to regulate loss of heat by the entire body. The dermis contains many sensory nerve endings. Sensations such as heat, cold, touch, and pressure are mediated by dermal nerve endings. There are two regions of the dermis:

☐ The **reticular layer** of the dermis is a thick region of irregularly arranged protein fibers. Most of the fibers are collagenous, but a few are made of elastin.

☐ The **papillary layer** is the bumpy superficial portion of the dermis attached to the epidermis. The bumps, called **papillae** (meaning "nipples"), form regular rows in thick skin but are rather irregularly arranged in thin skin. For this reason, thick skin can be observed to have distinct ridges, such as fingerprints. These ridges give the hands and feet better gripping ability.

☐ 3 Deep to the skin is a layer of **subcutaneous tissue**, sometimes called the **hypodermis** or **superficial fascia**. Although not a part of the skin, it is often studied along with skin. Subcutaneous tissue is loose, fibrous (areolar) connective tissue that connects the skin to underlying muscles and bone. Some of the areolar tissue has been modified to become adipose tissue. Adipose tissue's protective and insulating characteristics complement the protection and temperature regulation roles of the skin.

## *Study* TIPS

To remember the layers of the epidermis from the innermost to outermost layer, use the following mnemonics: **B**ragging **S**ure **G**ains **L**ittle **C**ooperation or **C**ome **L**et's **G**et **S**un **B**urned.

Hint ▶ Figure 10-1 shows light micrographs of skin structures.   •

## B. HAIR, NAILS, AND GLANDS

The skin has a variety of accessory structures, including **hair** and **nails**. Both hair and nails are modified forms of stratum corneum, or keratinized tissue. A hair is a cylinder of compact keratinized material, and a nail is a plate of compact keratinized material. Identify the structures described in the following in a model and in prepared specimens.

☐ 1 Each hair is formed within a separate **hair follicle**. The follicle is a sheathlike indentation of the epidermis (Figure 10-1, *D*). At the bottom of the follicle, a **hair papilla** covered with stratum germinativum produces the hair. The portion of each hair within the follicle is called the **hair root**, whereas the portion that has been pushed out of the follicle is called the **hair shaft**. The hair has a very dense cortex and a less dense medulla.

☐ 2 Attached to each follicle is an exocrine (ducted) sebaceous gland (see Figure 10-1, *D*). This gland produces the fatty substance, **sebum**, that coats the hair and skin. Sebum prevents moisture loss and conditions the hair and skin so that they do not become brittle and easily broken.

☐ 3 The **arrector pili** muscle is a strap of smooth muscle tissue connecting the side of a follicle to the superficial surface of the dermis. When contracted, the muscle pulls the follicle so that it is nearly perpendicular to the skin's surface (Figure 10-2). This increases the air spaces among the hairs, improving their insulation quality. Contraction of the arrector pili also dimples the epidermis, raising a ridge at the edge of the follicle (a "goose pimple" or "goose bump").

☐ 4 The toenail or fingernail is also formed by a modified portion of stratum germinativum. In the case of either hair or nail, this modified tissue is often called **matrix**. A portion of the **nail bed** (skin under the nail) is matrix that produces the nail plate. Part of the matrix may be visible through the nail as a pale crescent, or **lunula**. Nail formation begins under a fold of epidermis. The portion of the nail under the fold is the **root**, and the visible portion is the nail **body**. A **cuticle**, or **eponychium**, may extend from the fold onto the nail body (Figure 10-3).

☐ 5 **Sweat glands** are found in many areas of the skin. They are exocrine glands that produce a watery solution, sweat, that coats the skin. Sweat serves primarily to improve heat loss by the skin through evaporation. Eccrine sweat glands produce thin, watery sweat in many areas of the body. Apocrine sweat glands, found in the axillary and pubic regions, and starting at puberty, secrete a thicker sweat that is rich in complex organic molecules.

ⓘ *safety first*

Observe the usual precautions for dissection activities. Use gloves and take care to avoid injuries. Use safety goggles to avoid injury during dissections.   •

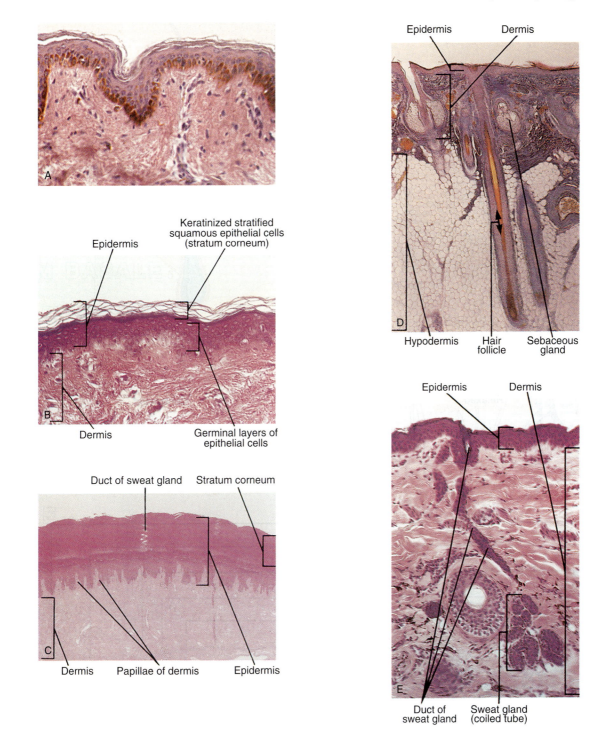

**FIGURE 10-1** Light micrographs of skin structures. **A** and **B**, Thin skin. **C**, Thick skin. **D**, Skin with hair follicles and other accessory structures. **E**, Skin with sweat glands.

## C. DISSECTION: CAT INTEGUMENT

If you are going to use the cat as a dissection specimen through-out this course, it must first be skinned. This exercise not only directs you on how to perform this essential step, but it also provides an excellent opportunity to practice identifying surface anatomy—as well as to see "the big picture" of the integumentary system **(Figure 10-4)**.

- ☐ 1 Examine the external aspect of your specimen **(Figure 10-5)**:
  - ☐ Determine the anatomical orientation of the specimen. Which direction is anterior? posterior? Which direction is ventral? dorsal? Identify sagittal, transverse, and frontal planes in your specimen.
  - ☐ Identify these externally visible features:
    - ☐ Pinna (auricle, or external ear)
    - ☐ External nares (nostrils)
    - ☐ Vibrissae (whiskers)
    - ☐ Integument (skin)
    - ☐ Forelimbs
    - ☐ Hindlimbs
    - ☐ Thoracic region
    - ☐ Abdominal region
    - ☐ Genitals (sex organs)
    - ☐ Anus
    - ☐ Tail
  - ☐ Determine the sex of your specimen by examining the external genitals **(Figure 10-6)**:
    - ☐ **Female**—Immediately anterior to the anus, on the ventral surface, is the **vulva** with an opening to the **vagina** and the **urethra**. The vulva is also called the urogenital opening.
    - ☐ **Male**—In the male, there is a pouch of skin immediately ventral to the anus. This pouch—the **scrotum**—contains the **testes**, or male primary sex organs. Anterior to the scrotum is the end of the **penis** with its **prepuce**, or skinfold covering. Locate the opening of the **urethra** in the penis.

- ☐ 2 Place the animal in the tray with its ventral surface facing you. Insert the tip of a scissors into the hole already present under the chin. This hole was used to inject the cat's vessels with latex. If you are using a cat that was not injected, then lift up a fold of skin and puncture it at the same site. Slide the bottom tip of the scissors into the **subcutaneous** area under the skin.

- ☐ 3 Begin cutting along the lines indicated in **Figure 10-5**. Notice that you must cut all the way around the neck and the distal ends of each limb (just proximal to the feet, or paws). All the other cuts are to be made only on the ventral aspect of the cat. When you make the median cut along the posterior abdomen, toward the tail, cut around the genitals rather than through them. Be careful not to cut into the skeletal muscles under the skin.

- ☐ 4 With your forceps, pull the flaps of skin over the neck away from the cat's body. Notice the **areolar tissue** under the skin that is pulled apart as you remove the skin. Sometimes it helps if you scrape at the loose connective tissue under the skin with your scalpel as you peel the skin away. The skinning process is not difficult if you have patience and proceed slowly. Peel the other flaps of skin in a similar fashion until the cat is completely skinned.

- ☐ 5 Examine the skin, identifying these features:
  - ☐ **Epidermis**—The thinner outer layer of the skin
  - ☐ **Dermis**—The thick inner layer of the skin
  - ☐ **Hypodermis**—The subcutaneous tissue under the skin proper, made of areolar and adipose tissue

- ☐ 6 Wrap the skin around the cat again and place the entire animal in its storage bag or other container provided.

> **Hint** ► After each dissection activity later in this course, always wrap the skin around your specimen before storing it. This will help prevent your specimen from drying out—a common problem with preserved specimens. •

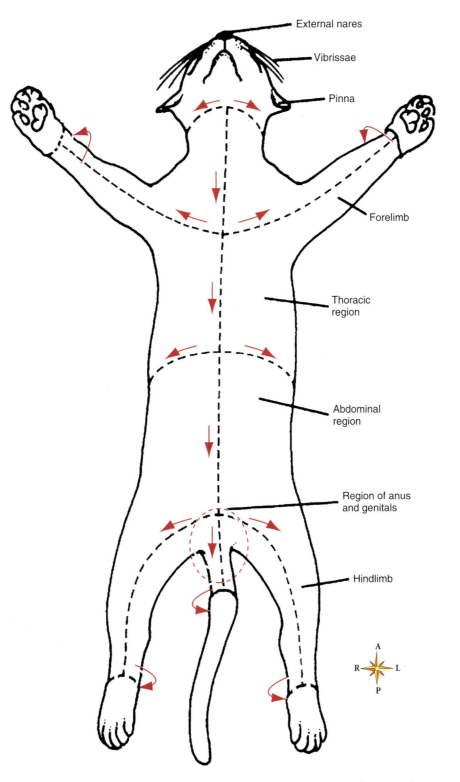

**FIGURE 10-5** Cat. External aspect, showing how skin is to be cut for removal.

## Coloring Exercises: Bone Tissue

Use colored pens or pencils to shade in both the figure and the labels. Each red numeral in the figure corresponds to a matching red numeral following the appropriate label.

COMPACT BONE 1
SPONGY BONE 2
BLOOD VESSEL 3
CANALICULI 4
CENTRAL CANAL 5
LACUNA 6
LAMELLA 7
MYELOID TISSUE 8
OSTEOCYTE 9
OSTEON 10

PERIOSTEUM 11
TRABECULA 12
TRANSVERSE CANAL 13
NERVE 14

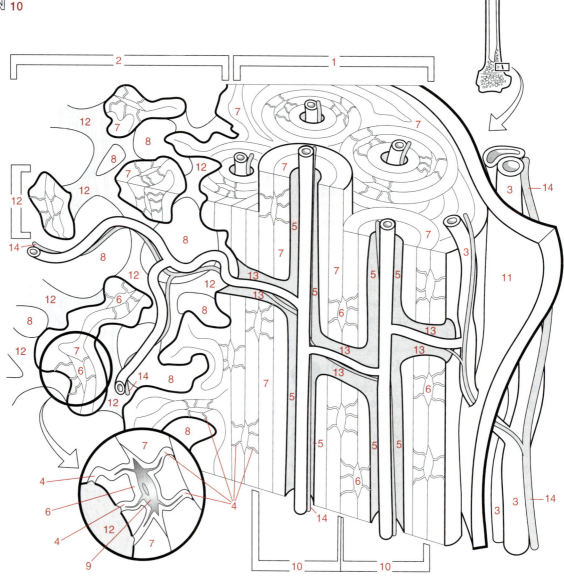

**FIGURE 11-2** Microscopic structure of bone.

FIGURE 11-3 Compact bone (ground bone) in cross-section. High power.

☐ 3 **Epiphyseal plate**—Until a long bone has stopped growing in length, a layer of cartilage called the **epiphyseal plate** remains between each epiphysis and the diaphysis (Figure 11-4). During periods of growth, proliferation of epiphyseal cartilage cells brings about a thickening of this layer. **Ossification** (bone formation) of the additional cartilage nearest the diaphysis then follows; that is, osteoblasts make new bone matrix. As a result of this process, the bone becomes longer.

> 📌 **landmark characteristics**
>
> The epiphyseal plate shown in Figure 11-4 is made up of four regions; each region, in turn, is made of several layers of cells. The layer closest to the epiphysis (top of figure) is not changing or growing, so the cells are said to be at rest. The *zone of proliferation*, which includes cells undergoing mitotic division, is where the plate becomes thicker. The *zone of hypertrophy* is where older, enlarged cells degenerate before calcification. In the *zone of calcification*, new cancellous bone can be seen. Typically, the epiphyseal plate is seen as a region of hyaline cartilage separated from a region of cancellous bone by a region of degenerating chondrocytes (cartilage cells) in lacunae that seem to be "stacked" in roughly parallel rows.  •

# D. THE PLAN OF THE SKELETON

The usual number given for bones in the human skeleton is 206. This is by no means the absolute exact number, however. Most people have more bones, but each person has different types, locations, and numbers of "extra" bones (Figure 11-5). Some people may be missing a bone or two. In this activity, you will examine both the standard 206 bones and the extra bones that may be present (Table 11-1).

☐ 1 Obtain an articulated (connected) human skeleton.
☐ 2 The standard **axial skeleton** consists of 80 bones that form the central axis of the skeleton (Figure 11-6). These 80 bones include 28 skull bones, 1 unattached bone in the throat, 26 vertebrae, and 25 rib cage bones. Locate the bones of the axial skeleton in your specimen. Do not worry

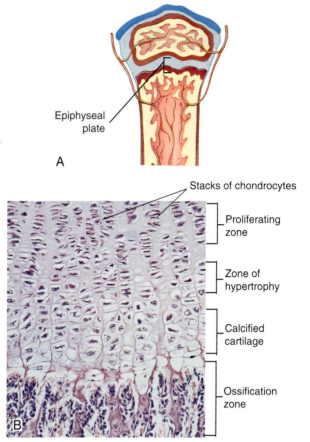

FIGURE 11-4 Epiphyseal plate. **A,** Localization of the epiphyseal plate between the epiphyses and diaphyses of a long bone. **B,** Zones of the epiphyseal plate.

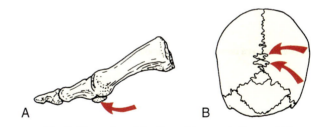

FIGURE 11-5 Examples of the "extra" bones. **A,** Sesamoid bone at the base of the thumb. **B,** Sutural (wormian) bones along a suture joint.

about learning the names of individual bones now. That will come later. For now, concentrate on "the big picture" of skeletal organization.

> ❗ **safety first**
>
> Be cautious when handling the articulated skeleton. The bones or mounting hardware may become loose and fall from the support frame, injuring you or your lab mates.  •

☐ 3 Locate the bones of the **appendicular skeleton** in your specimen. The appendicular skeleton, consisting of the 126 nonaxial bones, includes the bones of the appendages, or extremities (arms and legs). Sixty-four of these bones are in the *upper extremities* (shoulders and arms). Sixty-two bones are in the *lower extremities* (hips and legs).

☐ 4 Ask your lab instructor if there are any extra standard bones or any standard bones missing in your specimen. What difficulties could such differences have caused the individual during life?

☐ 5 Determine whether your specimen has any of the extra bone types typically found in skeletons (see Figure 11-5):

    ☐ **Sesamoid bones** are so called because they resemble sesame seeds: tiny rounded specks. Sesamoid bones are often found within tendons of the hand and foot.

    ☐ **Wormian bones**, also called *sutural bones*, are flat bones that form in the sutures (joints) between the cranial bones of the skull.

## E. BONE MARKING AND FEATURES

As you have already observed on your specimen, bones do not generally have a smooth surface. There are many bumps, holes, and projections on the bones of the human skeleton. These *bone markings* are named with terms that describe their shape and location. As a preview to the next few exercises, review the terms used to name bone markings listed in Table 11-2.

### Study TIPS

Bone markings may be difficult to identify. An effective strategy to identify the various bone markings would be to develop flash cards that have the bone marking on one side and a meaning or illustration/photo on the opposite side, or refer to *Mosby's Anatomy & Physiology Study and Review Cards*. Also, take advantage of the skeletal models in the lab, and with a partner, quiz each other on the various bone markings. Use your textbook, lab manual, the *A&P Survival Guide*, or atlas as a reference.

**TABLE 11-2**  **Bone Markings**

| MARKING | MEANING |
| --- | --- |
| Angle | A corner |
| Body | The main portion of a bone |
| Condyle | Rounded bump; usually fits into a fossa on another bone, forming a joint |
| Crest | Moderately raised ridge; generally a site for muscle attachment |
| Epicondyle | Bump near a condyle; often gives the appearance of a "bump on a bump"; for a muscle attachment |
| Facet | Flat surface that forms a joint with another facet or flat bone |
| Fissure | Long, cracklike hole for blood vessels and nerves |
| Foramen | Round hole for vessels and nerves (pl. *foramina*) |
| Fossa | Depression; often receives an articulating bone (pl. *fossae*) |
| Head | Distinct epiphysis on a long bone, separated from the shaft by a narrowed portion (or neck) |
| Line | Similar to a crest but not raised as much (is often rather faint) |
| Margin | Edge of a flat bone or flat portion of an irregular bone |
| Meatus | Tubelike opening or channel (pl. *meati*) |
| Neck | A narrowed portion, usually at the base of a head |
| Notch | A V-like depression in the margin or edge of a flat area |
| Process | A raised area or projection |
| Ramus | Curved portion of a bone, like a ram's horn (pl. *rami*) |
| Sinus | Cavity within a bone |
| Spine | Similar to a crest but raised more; a sharp, pointed process; for muscle attachment |
| Sulcus | Groove or elongated depression (pl. *sulci*) |
| Trochanter | Large bump for muscle attachment (larger than tubercle or tuberosity) |
| Tubercle | Smaller version of a tuberosity |
| Tuberosity | Oblong, raised bump, usually for muscle attachment |

Name: _____ Date: _____ Section: _____

## LAB REPORT 11
# Overview of the Skeleton

## Multiple Choice

1. _____
2. _____
3. _____
4. _____
5. _____
6. _____

## Figure 11-6 (from p. 115)

1. _____
2. _____
3. _____
4. _____
5. _____
6. _____
7. _____
8. _____
9. _____
10. _____
11. _____
12. _____
13. _____
14. _____
15. _____
16. _____
17. _____
18. _____
19. _____

## Multiple Choice (choose the one best response)

1. The inner lining of the medullary cavity is
   a. made of compact bone
   b. called the *endosteum*
   c. called the *periosteum*
   d. a and c are correct
2. Which of these tissues is present in a typical long bone?
   a. blood tissue
   b. cancellous bone
   c. compact bone
   d. dense fibrous tissue
   e. hyaline cartilage
   f. all of the above
3. In the coloring figure of the long bone, the epiphyseal plate is shown. What is the reason for its presence?
   a. It is scar tissue from a previous fracture.
   b. It is an area of growth between the epiphysis and diaphysis during bone development.
   c. It is callous tissue from overuse of the bone.
   d. It is the site of a current fracture.
4. The human skeleton functions to
   a. produce blood tissue
   b. store fat and minerals
   c. protect vital organs
   d. allow movement of the body
   e. provide a supporting framework
   f. all of the above
5. Your physician has just informed you that you have 32 bones in your skull. This means
   a. you have the standard number of skull bones
   b. you have some sesamoid bones in your skull
   c. you have some sutural bones in your skull
   d. you are missing some skull bones
6. The same physician tells you that all of your knee ligaments have been severed. This means that
   a. your leg bones are not being held together very well
   b. your leg muscles have become separated from the bone
   c. your patella (kneecap) is fractured
   d. your femur (thigh bone) is fractured

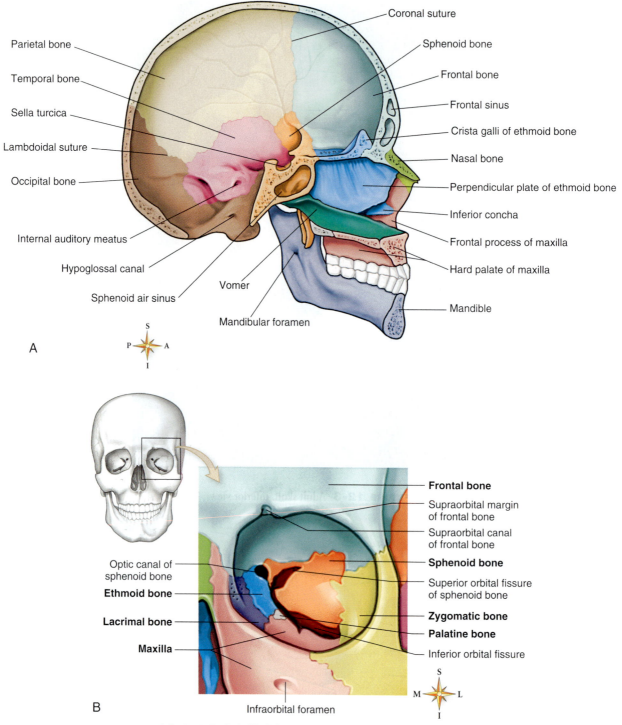

A, Right half of skull, medial view.

- Parietal bone
- Temporal bone
- Sella turcica
- Lambdoidal suture
- Occipital bone
- Internal auditory meatus
- Hypoglossal canal
- Sphenoid air sinus
- Vomer
- Mandibular foramen
- Coronal suture
- Sphenoid bone
- Frontal bone
- Frontal sinus
- Crista galli of ethmoid bone
- Nasal bone
- Perpendicular plate of ethmoid bone
- Inferior concha
- Frontal process of maxilla
- Hard palate of maxilla
- Mandible

B, Bones that form right orbit.

- Optic canal of sphenoid bone
- Ethmoid bone
- Lacrimal bone
- Maxilla
- Infraorbital foramen
- Frontal bone
- Supraorbital margin of frontal bone
- Supraorbital canal of frontal bone
- Sphenoid bone
- Superior orbital fissure of sphenoid bone
- Zygomatic bone
- Palatine bone
- Inferior orbital fissure

**FIGURE 12-4** **A,** Right half of skull, medial view. **B,** Bones that form right orbit.

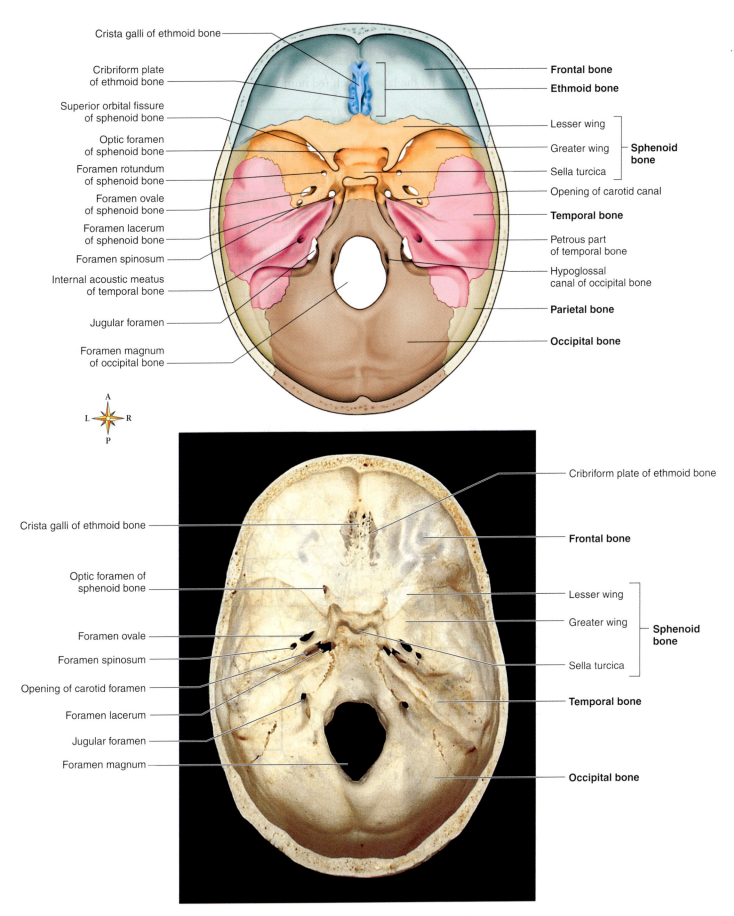

Crista galli of ethmoid bone

Cribriform plate of ethmoid bone

Superior orbital fissure of sphenoid bone

Optic foramen of sphenoid bone

Foramen rotundum of sphenoid bone

Foramen ovale of sphenoid bone

Foramen lacerum of sphenoid bone

Foramen spinosum

Internal acoustic meatus of temporal bone

Jugular foramen

Foramen magnum of occipital bone

**Frontal bone**
**Ethmoid bone**

Lesser wing

Greater wing    **Sphenoid bone**

Sella turcica

Opening of carotid canal

**Temporal bone**

Petrous part of temporal bone

Hypoglossal canal of occipital bone

**Parietal bone**

**Occipital bone**

Crista galli of ethmoid bone

Optic foramen of sphenoid bone

Foramen ovale

Foramen spinosum

Opening of carotid foramen

Foramen lacerum

Jugular foramen

Foramen magnum

Cribriform plate of ethmoid bone

**Frontal bone**

Lesser wing

Greater wing    **Sphenoid bone**

Sella turcica

**Temporal bone**

**Occipital bone**

**FIGURE 12-5** Adult skull, superior view showing floor of the cranial cavity.

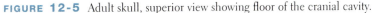

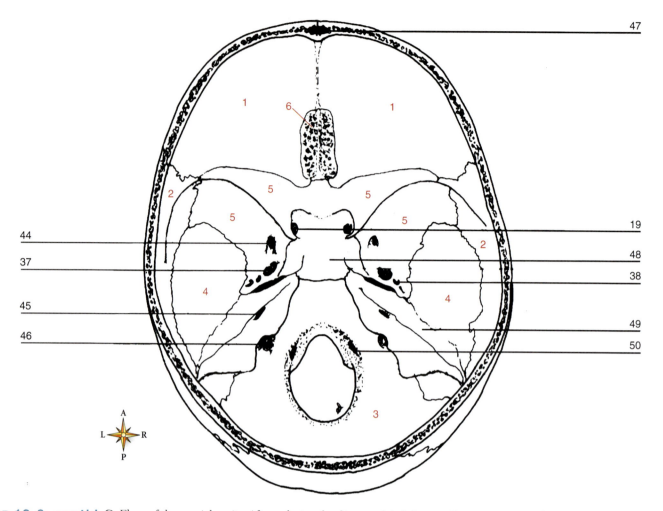

**FIGURE 12-6, cont'd** C, Floor of the cranial cavity. After coloring the diagram, label the specific structures noted on the lines provided and on the blanks in the Lab Report at the end of this exercise.

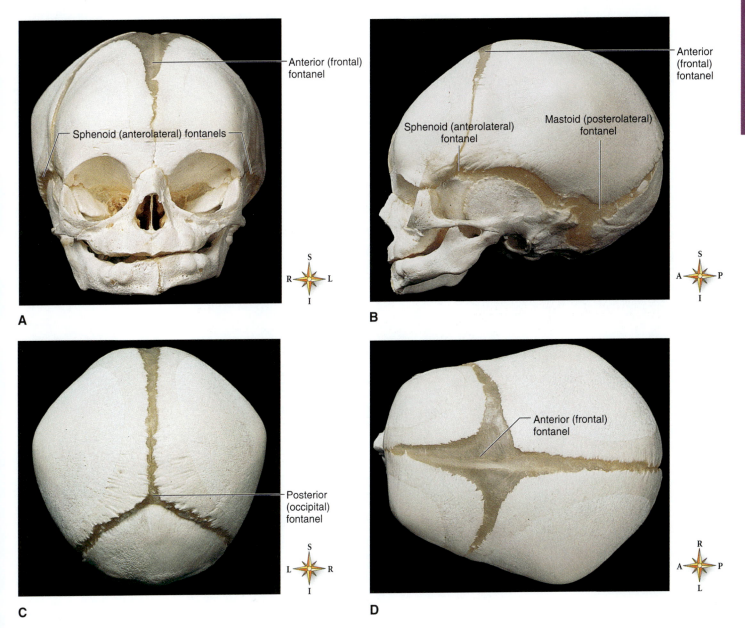

Anterior (frontal)
fontanel

Sphenoid (anterolateral) fontanels

Anterior
(frontal)
fontanel

Sphenoid (anterolateral)
fontanel

Mastoid (posterolateral)
fontanel

Posterior
(occipital)
fontanel

Anterior (frontal)
fontanel

A

B

C

D

**FIGURE 12-7**   The fetal skull **A,** Anterior view. **B,** Left lateral view. **C,** Posterior view. **D,** Superior view.

Name: _____ Date: _____ Section: _____

## LAB REPORT 12
# The Skull

**Figure 12-6** (from pp. 126-128)

(NOTE: Numbers 1 through 14 are used for coloring labels.)

15. _____
16. _____
17. _____
18. _____
19. _____
20. _____
21. _____
22. _____
23. _____
24. _____
25. _____
26. _____
27. _____
28. _____
29. _____
30. _____
31. _____
32. _____
33. _____
34. _____
35. _____
36. _____
37. _____
38. _____
39. _____
40. _____
41. _____
42. _____
43. _____
44. _____
45. _____
46. _____
47. _____
48. _____
49. _____
50. _____
51. _____

## List the Bones of the Cranium

1. _____
2. _____
3. _____
4. _____
5. _____
6. _____

## Name the Facial Bones

1. _____
2. _____
3. _____
4. _____
5. _____
6. _____
7. _____
8. _____

## Name the Auditory Ossicles

1. _____
2. _____
3. _____

## Name the Major Fontanels

1. _____
2. _____
3. _____
4. _____

## Sutures

1. _____
2. _____
3. _____
4. _____

## Markings

1. _____
2. _____
3. _____
4. _____
5. _____
6. _____
7. _____
8. _____
9. _____
10. _____

## Anatomical Relationships

1. _____
2. _____
3. _____
4. _____
5. _____
6. _____
7. _____
8. _____
9. _____
10. _____

## Sutures (determine the name of the suture joint described)

1. Joins the parietal bones together
2. Joins the superior margin of the temporal bone to the frontal, parietal, and occipital bones
3. Joins the palatine bone to the maxilla
4. Joins the frontal bone to the two parietal bones

## Markings (give the name of the bone marking or feature described in each item)

1. Large process of the temporal bone, just posterior to the external auditory meatus; it contains sinuses
2. A smaller, needle-shaped process just medial to the process described in the previous statement
3. A hole in the sphenoid bone that allows the optic nerve to exit the eye orbit
4. Same name for ridgelike processes on both the maxilla and mandible in which the teeth are embedded
5. A crest or projection on the superior surface of the ethmoid bone
6. A curved plate of bone projecting from the lateral wall of the nasal cavity, just above the inferior nasal conchae (part of the ethmoid bone)
7. One of two holes on the anterior portion of the mandibular body
8. A hole in the maxilla just below the orbit of the eye
9. The upper portion of the nasal septum is formed by this part of the ethmoid
10. This structure is formed by both the zygomatic process of the temporal bone and the temporal process of the zygomatic bone

## Anatomical Relationships (use a directional term to complete each item correctly)

1. The parietal bone is _?_ to the occipital bone.
2. The lacrimal bones are on the _?_ margin of the orbit.
3. The mandible is mostly _?_ to the maxilla.
4. The occipital condyles are _?_ to the foramen magnum.
5. The palatine bones are _?_ to the maxilla.
6. The superior conchae are _?_ to the nasal septum.
7. The mandibular condyle is _?_ to the coronoid process.
8. The parietal bones are _?_ to the frontal bone.
9. The frontal fontanel is _?_ to the posterior fontanel.
10. The incisive canal is _?_ to the transverse palatine suture.

# The Vertebral Column and Thoracic Cage

As you recall, the axial skeleton consists of the skull, the vertebral column, the thoracic cage, and the hyoid bone. All these structures form the central core, or *axis*, of the skeleton. In the previous exercise, we explored the skull in some detail. This exercise presents the remainder of the axial skeleton.

The **vertebral column** is a set of 26 bones stacked one on another to form a slightly curved, flexible support rod. The 25 bones of the rib cage, or **thoracic cage,** protect the lungs and heart within the thorax. The **hyoid** bone is not attached to any other bone but is very close to the skull and vertebral column and lies along the central axis of the body.

## BEFORE YOU BEGIN

- ☐ Read the appropriate chapter in your textbook.
- ☐ Set your learning goals. When you finish this exercise, you should be able to:
  - ☐ name the component bones of the vertebral column and thoracic cage
  - ☐ identify the bones of vertebral column and thoracic cage, and markings of these bones on a specimen and in figures
- ☐ Prepare your materials:
  - ☐ human skeleton (disarticulated)
  - ☐ human skeleton (articulated)
  - ☐ human vertebrae set (optional)
  - ☐ colored pencils or pens
  - ☐ demonstration pointer
- ☐ Read the directions and safety tips for this exercise *carefully* before starting any procedure.

> ⓘ *safety first*
>
> Be careful when handling skeletal specimens. Loose hardware can injure both you and the specimen. Do not forget to use only approved demonstration pointers. •

## *Study* TIPS

Because there are no distinct lines that separate the regions of the vertebra, the general features (lamina, pedicles, neural arch, processes, etc.) of each vertebra are often difficult to identify. You may want to obtain a photo or illustration of a vertebra and color/label each feature. It might be useful to do the same for the sacrum. Use your textbook, lab manual, the *A&P Survival Guide, Mosby's Anatomy & Physiology Study and Review Cards,* or *Brief Atlas* for reference.

> **Hint** ▶ Refer to **Figures 13-1** and **13-2** for color photographs of the bones and markings of the vertebral column. Figures in your textbook and *Brief Atlas* will also be useful. •

## A. THE VERTEBRAL COLUMN

The 26 bones of the vertebral column are grouped among the bones listed here. Find each bone and its important features on both an articulated skeleton and among the separate bones of a disarticulated skeleton or vertebrae set.

- ☐ 1 The seven **cervical** vertebrae at the superior end of the vertebral column are designated individually by number. The most superior is C1 (first cervical vertebra), the next is C2, and so on. Of C1 through C7, only the first two have commonly used alternate names:
  - ☐ **Atlas**—C1, or the atlas, is a ringlike vertebra that supports the skull by forming a joint with the occipital condyles. The atlas has an **anterior arch** and **posterior arch** fused to left and right **lateral masses** to form a circle. Flat articulating **facets** are found on both the superior and inferior aspects.
  - ☐ **Axis**—C2, or the axis, is remarkable for its **dens** or *odontoid* (toothlike) process. The dens points superiorly through the atlas to act as a pivot for the rotation of C1 and the skull.
  - ☐ In an articulated skeleton or figure, notice the **cervical curve** of the vertebral column produced by these seven vertebrae.
- ☐ 2 The 12 **thoracic** vertebrae (T1 through T12) are inferior to the cervical vertebrae. As a group, they curve in the opposite direction of the cervical curve to form the **thoracic curve.**
- ☐ 3 Five large **lumbar** vertebrae (L1 through L5) form the curve of the lower back, or **lumbar curve.**
- ☐ 4 All vertebrae from C3 to L5 have certain common features. Find each of these features on examples of all three vertebral types.
  - ☐ **Vertebral foramen**—Large hole (for the spinal cord) in the center
  - ☐ **Body**—Thick disk of bone at the anterior aspect of the vertebra and forms the anterior arch
  - ☐ **Pedicle**—Projects from the body and connects to lamina
  - ☐ **Lamina**—Thin, platelike section connecting to the pedicle anteriorly and from which processes project
  - ☐ **Superior articular process**—Projects superiorly from the lamina

# clinical applications: *anatomical images*

Perhaps the oldest method of producing images of internal structures of the human body is sketching dissected specimens. More recent technology has produced a number of advanced techniques of producing images of human structures. One such technique, first produced by Wilhelm Roentgen around the turn of the last century, is **radiography.** This technique, which is also called x-ray photography, uses radiation waves that pass through the specimen and onto photographic film. Because the waves pass more easily through soft tissue than through dense tissue, such as bone, shadows of some structures are visible when the film is developed.

**Figure 13-7, *A*** shows a radiograph, or x-ray photograph, of the lumbar region. Can you identify the structures indicated?

A more recent variation of this technique is computed tomography (CT). A CT scanner emits a beam of x-rays in a circular path around the subject. After passing through the body, the beams are processed by a computer that is able to reconstruct a three-dimensional image of internal structures. The operator can view different sections of the body on a video monitor. **Figure 13-7, *B*** shows a CT image of a horizontal section of the upper abdomen. A vertebra is clearly shown; can you identify its parts?

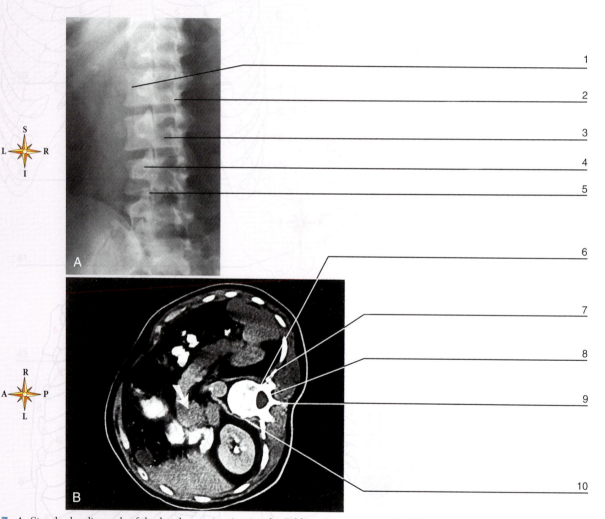

1
2
3
4
5
6
7
8
9
10

**FIGURE 13-7** **A,** Standard radiograph of the lumbar region (notice the "oblique" perspective). **B,** CT image of the abdomen. Label the structures noted on the lines provided and on the blanks in the Lab Report at the end of this exercise.

Name: _____ Date: _____ Section: _____

# The Vertebral Column and Thoracic Cage

**Figures 13-4 and 13-5** (from pp. 137-138)
(NOTE: Numbers 1 through 10 are used for coloring labels.)

11. _____
12. _____
13. _____
14. _____
15. _____
16. _____
17. _____
18. _____
19. _____
20. _____
21. _____
22. _____
23. _____
24. _____
25. _____
26. _____
27. _____
28. _____
29. _____
30. _____
31. _____
32. _____
33. _____
34. _____
35. _____
36. _____
37. _____
38. _____
39. _____
40. _____
41. _____
42. _____
43. _____
44. _____

**Table 13-1** (fill in the numbers as indicated)

**TABLE 13-1**    Count of Bones in the Vertebral Column of an Adult

| NUMBER | STRUCTURES |
|---|---|
| | Bones in the vertebral column |
| | Cervical vertebrae |
| | Thoracic vertebrae |
| | Lumbar vertebrae |
| | Spinal curves |
| | Bones in the thoracic cage |
| | Pairs of ribs |
| | Pairs of true ribs |
| | Pairs of false ribs |
| | Pairs of floating ribs |

**Figure 13-7** (from p. 140)

1. _____
2. _____
3. _____
4. _____
5. _____
6. _____
7. _____
8. _____
9. _____
10. _____

**Fill-in**

1. _____
2. _____
3. _____
4. _____
5. _____
6. _____
7. _____
8. _____
9. _____
10. _____
11. _____
12. _____
13. _____
14. _____
15. _____
16. _____
17. _____
18. _____
19. _____
20. _____

**Fill-in** (complete each statement with the appropriate term)

1. The first cervical vertebra is called C1, or _?_.
2. The second cervical vertebra is called C2, or _?_.
3. The inward curve of the lower back is called the _?_ curve.
4. All ribs articulate with the _?_.
5. Both radiographs and CT scans use _?_ to form images of anatomical structures.
6. Kyphosis is a condition in which the vertebral column is distorted, giving a person a humpback. This condition involves an exaggerated _?_ curve.
7. Both the coccyx and _?_ are vertebral bones that are actually fused vertebrae.
8. The thoracic vertebrae are _?_ to the lumbar vertebrae.
9. The superior portion of the sternum is called the _?_.
10. The only bone that does not articulate with another bone is the _?_.
11. The _?_ is a vertical tunnel through the sacrum for the passage of nerves descending from the spinal cord.
12. The medial bumps along the midline of your back are caused by the _?_ of each vertebra.
13. The _?_ ribs do not articulate, even indirectly, with the sternum.
14. The _?_ ribs connect directly to the sternum.
15. Costal cartilages are composed of _?_ cartilage tissue.
16. _?_ disks (appearing in Figure 13-2) form a type of joint between the bodies of vertebrae.
17. The _?_ is a horseshoe-shaped bone.
18. The tailbone is more properly known as the _?_.
19. Spinal nerves leave the protection of the vertebral column by way of lateral holes or gaps between the vertebrae called the _?_ foramina.
20. The medial ridge along the posterior surface of the sacrum is called the _?_.

# LAB EXERCISE 14
# The Upper Extremities

Remember that the human skeleton is divided into the *axial skeleton* and the *appendicular skeleton* (extremities). You are now ready to begin learning about the *upper* extremities.

The appendicular skeleton consists of all 126 bones that form the upper and lower extremities. The upper extremities have 64 bones altogether, including the bones of the shoulder girdle, arm, forearm, wrist, and hand. This exercise presents all the bones and many important bone features of the upper extremities.

## BEFORE YOU BEGIN

- ☐ Read the appropriate chapter in your textbook.
- ☐ Set your learning goals. When you finish this exercise, you should be able to:
  - ☐ name the bones of the upper extremity skeleton
  - ☐ identify upper extremity bones and markings on a specimen and in figures
- ☐ Prepare your materials:
  - ☐ human skeleton (disarticulated)
  - ☐ human skeleton (articulated)
  - ☐ colored pencils or pens
  - ☐ demonstration pointer
- ☐ Read the directions and safety tips for this exercise *carefully* before starting any procedure.

> **!** ***safety first***
> Beware of loose parts in the articulated skeleton. Remember to use only approved demonstration pointers. •

## *Study* TIPS

As discussed in Lab Exercise 12—The Skull, learning the bones will require harnessing your memorization capabilities. The more you practice, the better you will be at identifying the bones. It is extremely difficult to remember something when you are exposed to it only once. Remember this mantra: repeat it, repeat it, and repeat it! During the lab or during study time, take advantage of articulated and disarticulated skeletons, and with a partner, quiz each other over the different bones. Obtain unlabeled illustrations and use them to identify the bones. You can use your cell phone to photograph various bones and then print the photos and label them. There are many Internet sites that offer tutorials. An example of a very good interactive site is www.getbodysmart.com. This site has interactive links to all systems of the body. Use your text, lab manual, *A&P Survival Guide*, and *Mosby's Anatomy & Physiology Study and Review Cards* for reference.

> **Hint** Refer to **Figures 14-1** and **14-2** for color photographs of the bones and markings of the upper extremities. Figures in your textbook and *Brief Atlas* will also be useful. •

## A. THE SHOULDER GIRDLE

Find each bone and marking listed in a specimen and in figures.

- ☐ 1 The **scapula,** or shoulder blade, forms part of the shoulder girdle that supports the arms. It is an irregular bone on the superior, posterior aspect of the rib cage that extends laterally to form a joint with the arm. Find these features of the scapula:
  - ☐ **Axillary border**—Edge of the triangular scapula that faces the armpit region
  - ☐ **Vertebral border**—Scapular edge that faces the vertebral column
  - ☐ **Superior border**—Superior edge of the scapula
  - ☐ **Spine**—Large dorsal crest extending from the vertebral border to the acromion process
  - ☐ **Acromion process**—Large process projecting laterally from the scapular spine to form part of the shoulder joint
  - ☐ **Coracoid process**—Process anterior to, but smaller than, the acromion process
  - ☐ **Glenoid cavity**—Depression at the upper, lateral corner of the scapular triangle; it forms part of the shoulder socket
- ☐ 2 The **clavicle,** or collarbone, is the other bone of the shoulder girdle. The clavicle is a long bone whose long axis lies along a horizontal axis on the anterior, superior aspect of the thoracic cage.

## B. THE ARM AND FOREARM

Identify the bones and markings listed.

- ☐ 1 The **humerus** is the large, long bone of the arm. Find these parts of the humerus:
  - ☐ **Head**—Enlarged, rounded proximal portion
  - ☐ **Anatomical neck**—Diagonal line just inferior to the articular surface of the head
  - ☐ **Surgical neck**—The point at which the bone narrows significantly distal to the head
  - ☐ **Greater tubercle**—Large bump near the lateral aspect of the head
  - ☐ **Lesser tubercle**—Bump near the midline of the head's anterior aspect
  - ☐ **Intertubercular sulcus**—Groove between the greater and lesser tubercles

- ☐ **Deltoid tuberosity**—Rough muscle attachment site near the middle of the lateral surface of the shaft
- ☐ **Coronoid fossa**—Depression on the anterior, distal surface (between the epicondyles) that articulates with the ulna's coronoid process
- ☐ **Olecranon fossa**—Depression on the posterior, distal surface that articulates with the ulna's olecranon
- ☐ **Trochlea**—Spool-shaped articulating surface on the medial aspect of the distal end that articulates with the ulna's trochlear notch
- ☐ **Capitulum**—Ball-like articulating surface on the distal end, lateral to the trochlea, and articulates with the head of the radius
- ☐ **Medial epicondyle**—Blunt projection of bone medial to the trochlea
- ☐ **Lateral epicondyle**—Blunt projection of bone lateral to the capitulum

☐ 2 The **radius** is the lateral bone of the two bones of the forearm and articulates with the capitulum of the humerus. Find these features:
   - ☐ **Head**—Proximal disklike enlargement
   - ☐ **Radial tuberosity**—Oval bump just distal to the head, on the medial surface (facing the ulna)
   - ☐ **Styloid process**—Pointed bump on the lateral aspect of the distal end

☐ 3 The **ulna**, medial to the radius, articulates with the trochlea of the humerus. Find these markings of the ulna:
   - ☐ **Olecranon**—Large proximal process that forms the blunt point on the posterior aspect of the flexed elbow joint

- ☐ **Trochlear (semilunar) notch**—Curved anterior indentation on the olecranon that articulates with the trochlea of the humerus
- ☐ **Coronoid process**—Sharp anterior lip of the trochlear notch that articulates with the coronoid fossa of the humerus
- ☐ **Head**—Distal enlargement
- ☐ **Styloid process**—Pointed bump on the medial aspect of the head

## C. THE WRIST AND HAND

Identify the bones and markings listed.

☐ 1 The **carpus**, or wrist, is composed of eight small bones:
   - ☐ Pisiform
   - ☐ Triquetrum (triangular)
   - ☐ Lunate
   - ☐ Scaphoid (navicular)
   - ☐ Hamate
   - ☐ Capitate
   - ☐ Trapezoid
   - ☐ Trapezium

☐ 2 The hand consists of five similar **metacarpal** bones, numbered one through five (starting from the thumb side). Articulating with these are the **phalanges**, or finger bones. The phalanges are named *proximal*, *middle*, and *distal* and numbered one through five, according to their position. The first digit (pollex) does not have a middle phalanx.

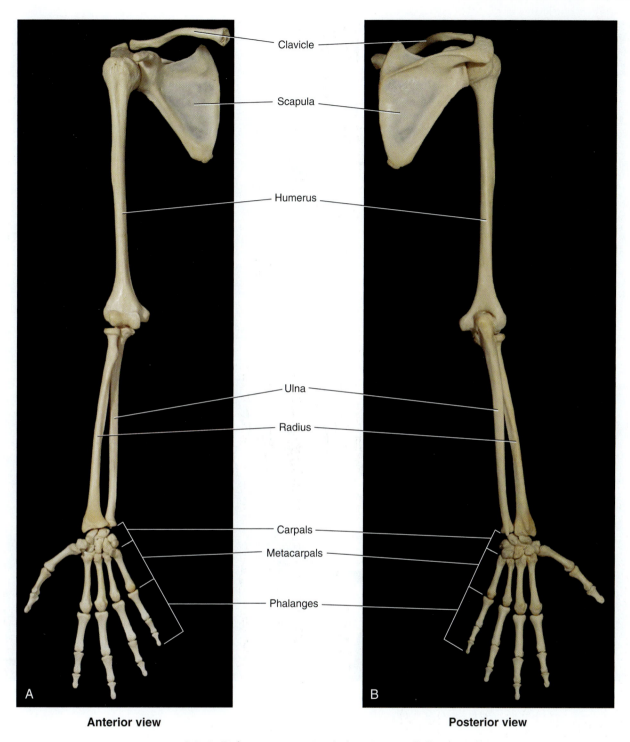

Clavicle

Scapula

Humerus

Ulna

Radius

Carpals

Metacarpals

Phalanges

A

B

**Anterior view**

**Posterior view**

**FIGURE 14-1** Right upper extremity. **A,** Anterior view. **B,** Posterior view.

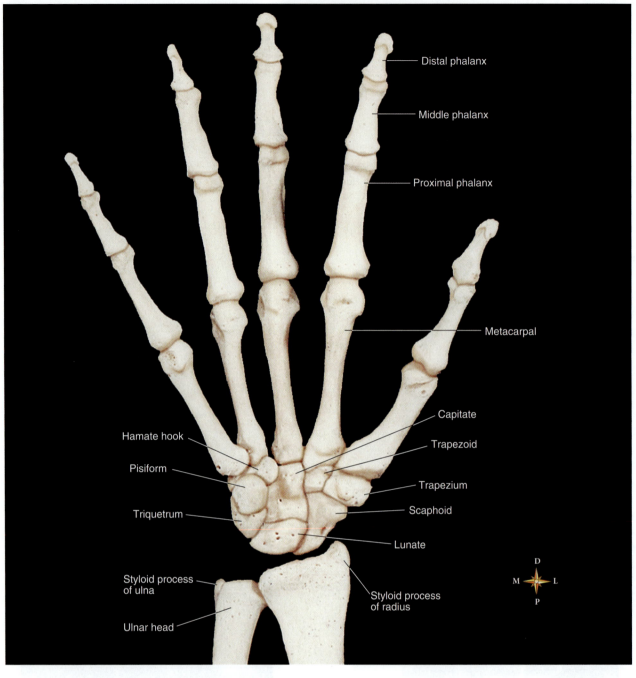

**FIGURE 14-2** Bones of the hand, palmar view.

## Coloring Exercises: Upper Extremity

Use colored pens or pencils to shade in both the figures and the labels. Each red numeral in the figure corresponds to a matching red numeral following the appropriate label.

SCAPULA 1
CLAVICLE 2
HUMERUS 3
RADIUS 4
ULNA 5
PISIFORM 6
TRIQUETRUM 7
LUNATE 8
SCAPHOID 9

HAMATE 10
CAPITATE 11
TRAPEZOID 12
TRAPEZIUM 13
METACARPAL 14
PROXIMAL PHALANX 15
MIDDLE PHALANX 16
DISTAL PHALANX 17

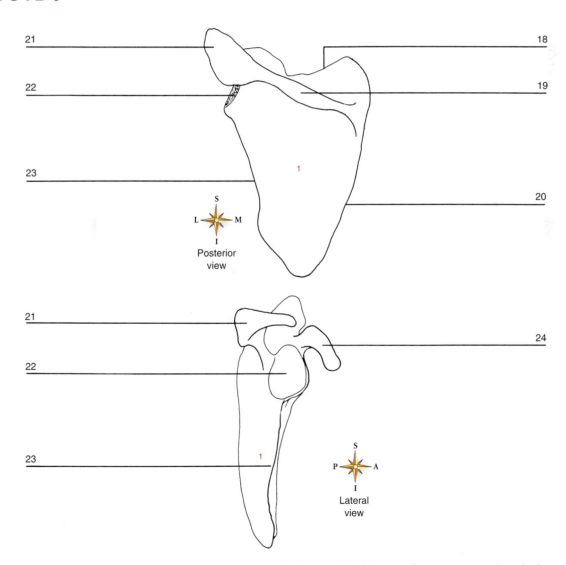

**FIGURE 14-3** Proximal structures of the upper extremity. After coloring the diagram, label the specific structures noted on the lines provided and on the blanks in the Lab Report at the end of this exercise.

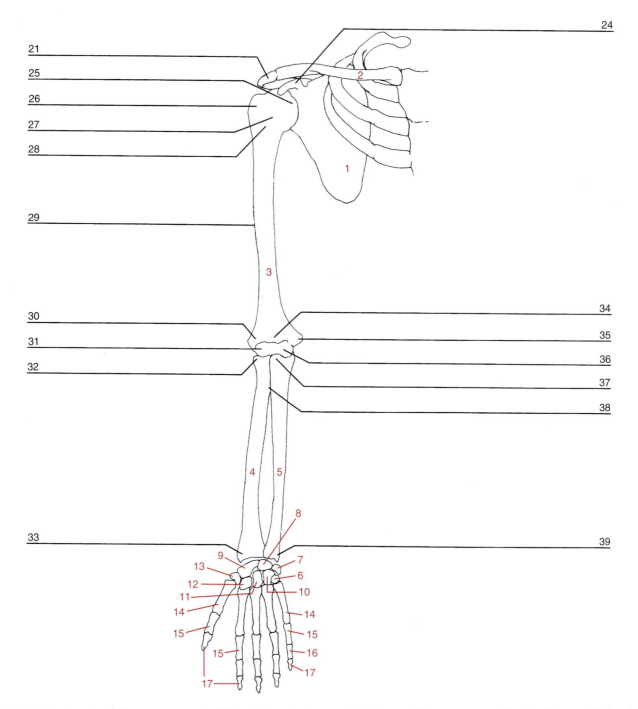

**FIGURE 14-4** Anterior view of the upper extremity. After coloring the diagram, label the specific structures noted on the lines provided and on the blanks in the Lab Report at the end of this exercise.

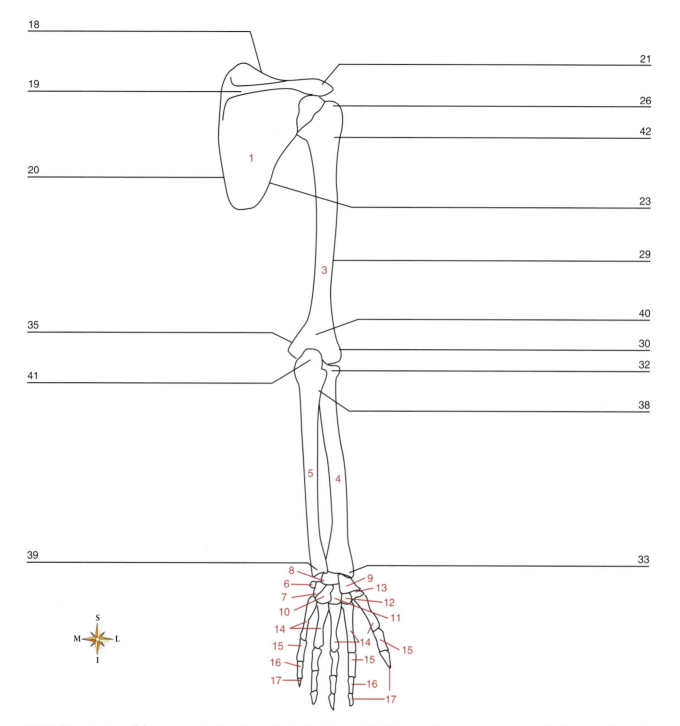

18

19

20

35

41

39

21

26

42

23

29

40

30

32

38

33

S
M    L
I

**FIGURE 14-5** Posterior view of the upper extremity. After coloring the diagram, label the specific structures noted on the lines provided and on the blanks in the Lab Report at the end of this exercise.

Name: _____ Date: _____ Section: _____

**LAB REPORT 14**
# The Upper Extremities

**Figures 14-3 through 14-5** (from pp. 147-149)
(NOTE: Numbers 1 through 17 are used for coloring labels.)

18. _____
19. _____
20. _____
21. _____
22. _____
23. _____
24. _____
25. _____
26. _____
27. _____
28. _____
29. _____
30. _____
31. _____
32. _____
33. _____
34. _____
35. _____
36. _____
37. _____
38. _____
39. _____
40. _____
41. _____
42. _____

**Multiple Choice**

1. _____
2. _____
3. _____
4. _____
5. _____
6. _____

**Multiple Choice** (only one response is correct in each item)

1. Jeb accidentally smashed his hand in a car door, breaking one of his bones. If the injury is near the middle of his palm, which type of bone must be involved?
   a. proximal phalanx
   b. metacarpal
   c. metatarsal
   d. distal phalanx
   e. carpal
2. The humeral process that articulates with the ulna is the
   a. trochlea
   b. capitulum
   c. lateral epicondyle
   d. medial epicondyle
   e. a and b are correct
3. The largest long bone of the upper extremities is the
   a. humerus
   b. tibia
   c. femur
   d. ulna
   e. clavicle
4. The forearm contains the
   a. ulna
   b. tibia
   c. radius
   d. humerus
   e. a and b are correct
   f. a and c are correct
5. The large, proximal process of the ulna that forms the posterior "point" of the flexed elbow is the
   a. head
   b. styloid process
   c. trochlea
   d. olecranon
   e. radial tuberosity
6. While skiing in the Alps last winter, Laura fell and fractured her collarbone. To which bone(s) does this collarbone attach?
   a. scapula
   b. sternum
   c. floating rib
   d. a and b are correct
   e. a, b, and c are correct

## Table 14-1  (fill in the numbers as indicated)

**TABLE  14-1**   Bone Count Related to the Upper Extremities

| NUMBER | STRUCTURES |
|--------|------------|
|        | Bones in the appendicular skeleton |
|        | Bones in the upper extremities |
|        | Wrist bones (total—right and left) |
|        | Hand/finger bones (total—right and left) |

# LAB EXERCISE 15
# The Lower Extremities

Recall that the appendicular skeleton consists of all 126 bones that form the upper and lower extremities. The lower extremities have 62 bones altogether, including the bones of the pelvic girdle, thigh, leg, ankle, and foot. This exercise presents all the bones and many important bone features of the lower extremities.

## BEFORE YOU BEGIN

- ☐ Read the appropriate chapter in your textbook.
- ☐ Set your learning goals. When you finish this exercise, you should be able to:
  - ☐ name the bones of the lower extremity
  - ☐ identify lower extremity bones and markings on a specimen and in figures
- ☐ Prepare your materials:
  - ☐ human skeleton (disarticulated)
  - ☐ human skeleton (articulated)
  - ☐ colored pencils or pens
  - ☐ demonstration pointer
- ☐ Read the directions and safety tips for this exercise *carefully* before starting any procedure.

> **⚠ safety first**
> Beware of loose parts in the articulated skeleton. Remember to use only approved demonstration pointers. •

### *Study* TIPS

As discussed in Lab Exercise 12—The Skull and Lab Exercise 14—The Upper Extremities, learning the bones will require harnessing your memorization capabilities. The more you practice, the better you will be at identifying the bones. It is extremely difficult to remember something when you are exposed to it only once. Remember this mantra: repeat it, repeat it, and repeat it! During the lab or during study time, take advantage of articulated and disarticulated skeletons, and with a partner, quiz each other over the different bones. Obtain unlabeled illustrations and use them to identify the bones. You can use your cell phone to photograph various bones and then print the photos and label them. There are many Internet sites that offer tutorials. An example of a very good interactive site is www.getbodysmart.com. This site has interactive links to all systems of the body. Use your text, lab manual, the *Brief Atlas,* the *A&P Survival Guide,* and *Mosby's Anatomy & Physiology Study and Review Cards* for reference.

 **Hint** ▶ Refer to **Figures 15-1 through 15-3** for color photographs of the bones and markings of the lower extremities. Figures in your textbook and *Brief Atlas* will also be useful. •

## A. THE PELVIC GIRDLE

Find the bones and markings listed.

- ☐ 1 A left and right **coxal bone,** or pelvic bone, join to form the pelvic girdle. The *coxae* articulate with the sacrum and support the legs. The coxae are formed by the fusion of three bones early in development: the **ilium, ischium,** and **pubis.** These bones are still identifiable as pelvic regions:
  - ☐ **Ilium**—Lateral, superior region of the bone; it includes the bowl-like wall of the upper pelvis
  - ☐ **Ischium**—Inferior, posterior, lateral coxal region
  - ☐ **Pubis**—Inferior, anterior, medial region, consisting of an upper and lower *ramus*; forms the pubic arch at the body's anterior midline
- ☐ 2 Find these other features of the coxal bone:
  - ☐ **Acetabulum**—Cuplike socket formed where the ilium, ischium, and pubis articulate with the head of the femur
  - ☐ **Iliac crest**—Upper, curving "lip" of the "bowl" formed by the iliac portion
  - ☐ **Iliac spines**—The "corners" of the iliac portion:
    - ☐ **Anterior superior iliac spine**—Blunt anterior point at the end of the iliac crest
    - ☐ **Anterior inferior iliac spine**—Smaller blunt point a short distance inferior to the anterior superior iliac spine
    - ☐ **Posterior superior iliac spine**—Point at the posterior end of the iliac crest
    - ☐ **Posterior inferior iliac spine**—Point a short distance inferior to the posterior superior iliac spine
  - ☐ **Ischial spine**—Pointed process on the inferior edge of the ischial region that protrudes into the pelvic outlet
  - ☐ **Obturator foramen**—Large hole formed by the curve of the pubis (medially) and the ischium
  - ☐ **Pubic arch**—Arch formed by both the left and right pubis's inferior angles, or rami
  - ☐ **Pelvic brim (inlet)**—Boundary of the opening into the **true pelvis**; formed by the upper margins (**pubic crests**) of the superior pubic rami, the *iliopectineal lines* that run along the inside face of the ilium, and the *sacral promontory*
  - ☐ **True pelvis**—Also called the *pelvis minor*, it is the space inferior to the pelvic brim in which the pelvic organs usually lie
  - ☐ **False pelvis**—Also called the *pelvis major*, it is the broad, shallow space superior to the pelvic brim and inferior to the iliac crests (actually part of the abdominal cavity)

☐ 3 Each coxal bone forms a joint directly with two other bones: the opposite coxal bone (anteriorly) and the sacrum (posteriorly).
  ☐ **Symphysis pubis**—Joint between the left and right pubis; also called the *pubic symphysis*
  ☐ **Sacroiliac joint**—Joint between the coxal's ilium and the lateral surface of the sacrum

## B. THE THIGH AND LEG

Find the bones and markings listed.

☐ 1 The **femur** is the long bone of the thigh. Find these features of the femur:
  ☐ **Head**—Large spherical enlargement at the proximal end
  ☐ **Neck**—Narrow region just distal to the head
  ☐ **Greater trochanter**—Large bump that forms a lateral, proximal "corner"
  ☐ **Lesser trochanter**—Bump on the medial, proximal aspect (just distal to the neck)
  ☐ **Intertrochanteric line**—Line that runs between the greater and lesser trochanter
  ☐ **Linea aspera**—Ridge that runs lengthwise along the posterior surface of the shaft
  ☐ **Medial condyle**—Medial, distal bump that articulates with the tibia
  ☐ **Lateral condyle**—Lateral, distal bump that articulates with the tibia
  ☐ **Medial epicondyle**—Blunt projection from the side of the medial condyle
  ☐ **Lateral epicondyle**—Blunt projection from the side of the lateral condyle
☐ 2 The **patella** is a large sesamoid bone forming the anterior bone of the knee joint.

☐ 3 The **tibia** is one of two long bones of the leg. The tibia is the larger, medial leg bone. Find these tibial features:
  ☐ **Lateral condyle**—Lateral, proximal articular surface
  ☐ **Medial condyle**—Medial, proximal articular surface
  ☐ **Crest**—Sharp ridge running longitudinally along the anterior surface
  ☐ **Tibial tuberosity**—Rough bump on the anterior aspect, just distal to the condyles
  ☐ **Medial malleolus**—Pointed bump on the medial aspect of the distal end
☐ 4 The **fibula** is the narrower, lateral bone of the leg, having these parts (Figure 15-1):
  ☐ **Head**—Proximal enlargement
  ☐ **Lateral malleolus**—Pointed bump on the lateral aspect of the distal end

## C. THE ANKLE AND FOOT

Find the bones and markings listed.

☐ 1 The seven **tarsal** bones form the ankle:
  ☐ **Talus**
  ☐ **Calcaneus**
  ☐ **Cuboid**
  ☐ **Navicular**
  ☐ **Medial cuneiform**
  ☐ **Intermediate cuneiform**
  ☐ **Lateral cuneiform**
☐ 2 The foot comprises five **metatarsal bones**, similar to the metacarpals of the hand. Also like the hand, the foot has 28 **phalanges** and the first digit (hallux) does not have a middle phalanx.

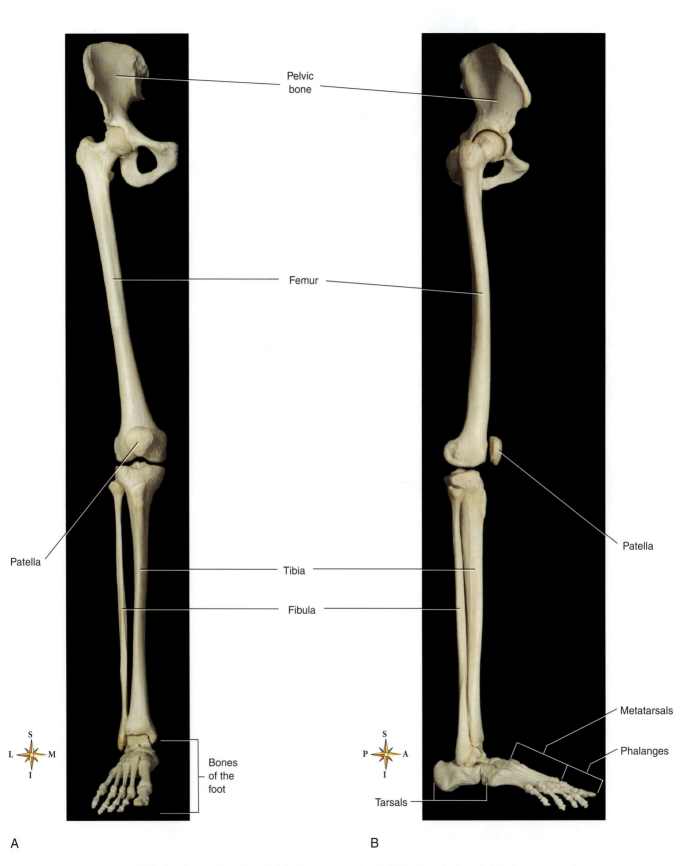

Pelvic
bone

Femur

Patella

Patella

Tibia

Fibula

Bones
of the
foot

Metatarsals

Phalanges

Tarsals

A

B

**FIGURE  15-1**  **A,** Anterior view of right lower extremity. **B,** Right lateral view of right lower extremity.

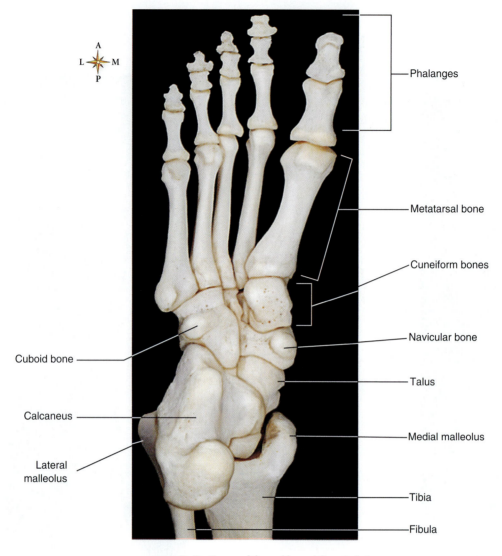

**FIGURE 15-2** Bones of the ankle and foot. Inferior aspect.

# clinical applications: *radiograph of the foot*

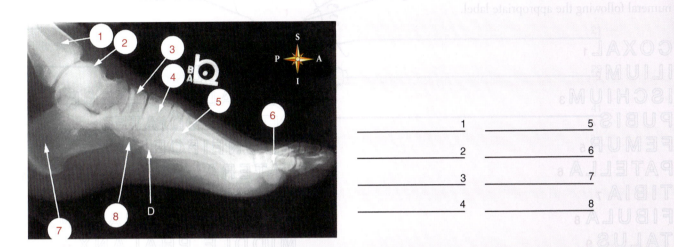

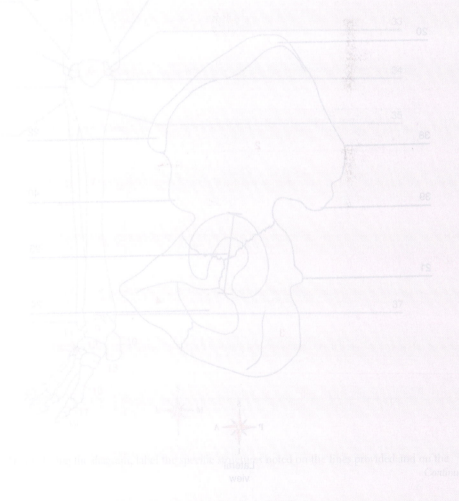

| | |
|---|---|
| 1 _____ | 5 _____ |
| 2 _____ | 6 _____ |
| 3 _____ | 7 _____ |
| 4 _____ | 8 _____ |

**FIGURE 15-3** Radiograph of the foot. This radiographic, or x-ray, image is a medial view of the left foot. Try to identify the noted structures and label them on the lines provided and on the blanks on the Lab Report at the end of this exercise.

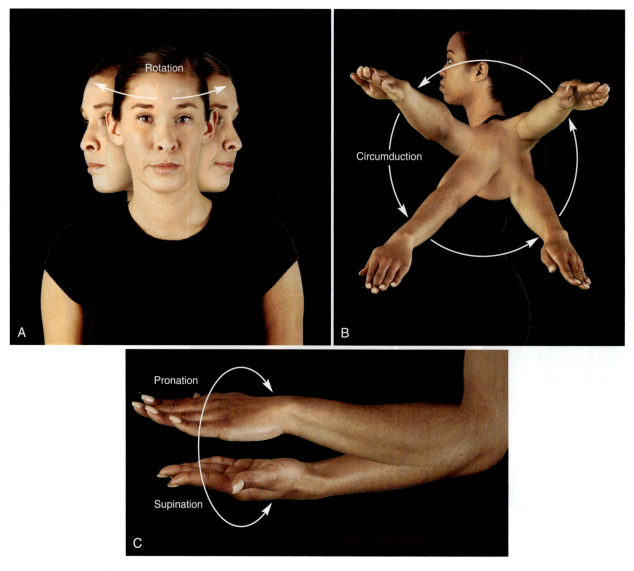

**FIGURE 16-6**  Circular movements. **A,** Rotation of the head. **B,** Circumduction of the arm. **C,** Pronation and supination of the hand.

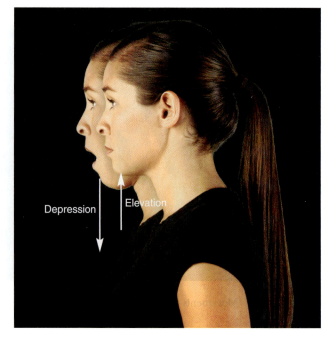

**FIGURE 16-7**  Special movements, temporomandibular joint (TMJ).

# clinical application: *arthrography*

Radiographic examination of the soft tissues that form joints is called **arthrography**. In one method of arthrography, a **contrast medium** that absorbs x-rays is injected into the joint. The liquid contrast medium coats the surfaces of the cartilage, ligaments, and other soft structures of the joint. When a regular radiograph is taken, the coated soft tissues clearly appear because the x-rays do not penetrate the coating. The resulting image is called an **arthrogram**.

In the knee arthrograms shown in **Figure 16-8**, several features that are invisible in a regular radiograph can be seen:

☐ **Bursa**—A "pillow" made of synovial membrane and filled with synovial fluid

☐ **Meniscus**—A piece of cartilage shaped like a disk with a curved surface

☐ **Articular cartilage**—A cartilage coating on the articulating surfaces of bone

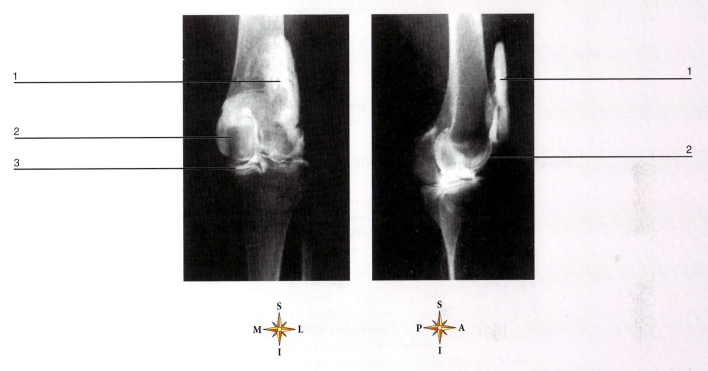

**FIGURE 16-8** Arthrograms of the knee joint. Label the structures noted on the lines provided and answer the questions below. Also, provide the labeling answers on the blanks in the Lab Report at the end of this exercise.

**1.** Can you identify these three joint structures in the arthrograms above?

**2.** What advantages does using contrast medium give when doing radiography?

**3.** Computed tomography (CT) has been more recently used to study joints. What advantages does the CT method have over the method described above?

## Identify

1. _____

2. _____

3. _____

4. _____

5. _____

6. _____

7. _____

8. _____

9. _____

10. _____

**Identify** (in the blanks to the left, write the type of motion illustrated in the matching figure to the right. For example, in number 1, is abduction or adduction shown?)

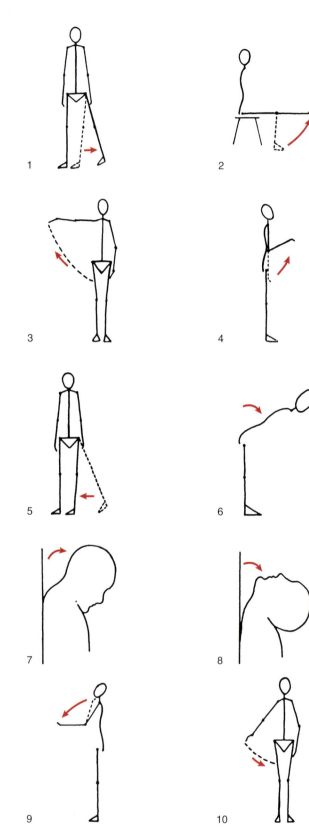

# Organization of the Muscular System

The muscular system is composed of all the skeletal muscles, sometimes called *muscle organs*, to distinguish them from muscle tissue and muscle cells. Skeletal muscles are attached to the skeleton in ways that move skeletal parts when the muscles contract. The chief function of this system, then, is skeletal movement. Secondarily, this system helps in temperature regulation by providing a source of metabolic heat.

This exercise presents some preliminary information about the muscular system. Although we have seen some of the histology of skeletal muscle tissue, we have not yet explored the muscle cell or the muscle organ. In this exercise we do both. We also take a brief look at how muscle organs are named so that the next exercise, which involves muscle identification, will be easier.

## BEFORE YOU BEGIN

☐ Read the appropriate chapter in your textbook.
☐ Set your learning goals. When you finish this exercise, you should be able to:
   ☐ describe the structure of the skeletal muscle cell
   ☐ identify and describe the principal structures of a skeletal muscle organ
   ☐ define the terms *origin* and *insertion*
   ☐ interpret the meaning of muscle names
☐ Prepare your materials:
   ☐ model or chart of a skeletal muscle cell
   ☐ model or chart of a skeletal muscle organ
   ☐ animal muscle specimen (fresh)
   ☐ dissection tools and tray
   ☐ protective gear: gloves, safety eyewear, and apron
   ☐ colored pencils or pens
☐ Read the directions and safety tips for this exercise *carefully* before starting any procedure.

## A. THE SKELETAL MUSCLE CELL

Skeletal muscle cells provide the basic functional capability of the muscle organ because they can contract with great force. Such contraction requires a rather unusual structural pattern in muscle cells if it is to be efficient. Study these features of the skeletal muscle cell in a model or figure **(Figures 17-1 through 17-3)**:

☐ 1  The **sarcolemma** is the plasma membrane of the long, cylindrical **muscle fiber,** or muscle cell. The sarcolemma has a *resting potential*, or electrical charge. This charge temporarily reverses during an *action potential*, or impulse, when the muscle fiber is stimulated.

☐ 2  The sarcolemma dips inward at several points to form internal *transverse tubules* called **T tubules.** An impulse traveling along the sarcolemma can thus also travel inside the cell.

☐ 3  Inside the cell are a number of membranous networks similar to endoplasmic reticulum (ER). They are all part of the **sarcoplasmic reticulum (SR).** The SR receives the impulse from the T tubules, which are nearby, and releases calcium ions into the **sarcoplasm** (cytoplasm of the muscle fiber) in response.

☐ 4  The calcium ions released from the SR diffuse through the sarcoplasm among parallel bundles of protein **myofilaments.** Each bundle is called a **myofibril** and is composed of an orderly arrangement of myofilaments called **thin filaments** and **thick filaments.**

☐ 5  The myofilaments are arranged in a repeating pattern and each is called a **sarcomere** (the basic contractile unit of a muscle fiber). When calcium ions react with some of the myofilament molecules, they slide past one another, shortening each sarcomere. The more sacromeres that shorten, the more a cell shortens.

☐ 6  As you recall, a muscle fiber has multiple **nuclei.** The nuclei are against the inside of the sarcolemma.

Use colored pens or pencils to shade in both the figure and the labels. Each red numeral in the figure corresponds to a matching red numeral following the appropriate label.

SARCOPLASMIC
  RETICULUM 1
NUCLEUS 2
SARCOLEMMA 3

BLOOD CAPILLARY 4
MYOFIBRIL 5
MITOCHONDRION 6
TRANSVERSE TUBULE 7

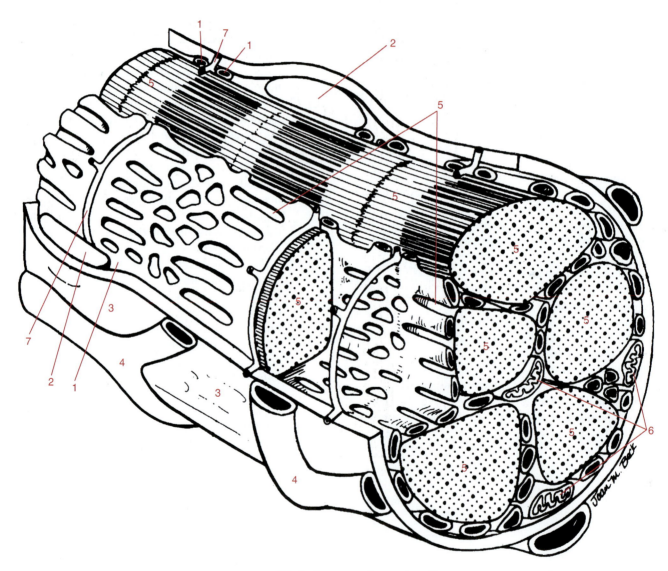

FIGURE 17-1 Structure of a muscle cell.

# B. STRUCTURE OF A MUSCLE ORGAN

Each muscle organ has a unique shape and size, but the principal components of each are the same (see Figure 17-2). Identify each of these in a model or chart. Some structures may be visible in a fresh muscle specimen.

> ⚠ **safety first**
> Use gloved hands when handling a fresh specimen at room temperature. You may wish to use dissection tools. If so, be careful to avoid cuts and puncture wounds. For example, use a scalpel only when necessary, and then, put it down in a safe place. Do not use it as a pointer. Many accidents have happened when one student points with a scalpel and another uses his/her finger, resulting in the scalpel cutting his/her finger.

☐ 1 The entire organ is covered with fibrous connective tissue that forms a sheath called the **epimysium**.

☐ 2 The fibrous tissue of the epimysium extends inward to form sheaths around bundles, or **fasciculi**, of muscle cells. This inner fibrous sheath is called the **perimysium**.

☐ 3 The perimysium's fibrous tissue continues inside each fascicle to wrap around each individual muscle cell. The connective tissue sheath around individual muscle fibers is called the **endomysium**. Just deep to the endomysium is the sarcolemma.

☐ 4 All the fibrous sheaths are continuous with one another and contain blood vessels that supply the muscle fibers. At the ends of the muscle organ, the muscle fibers stop, while the connective sheaths continue. This forms an extension of dense fibrous connective tissue that attaches to the periosteum of a bone (or to another muscle). If the fibrous connection is in the shape of a strap or band, it is called a **tendon.** If it is a broad, flat sheet, it is called an **aponeurosis**.

☐ 5 Generally a muscle attaches to the skeleton at two ends. The end that attaches to the more stationary bone is called the **origin**. The other end, or **insertion**, attaches to the bone that moves as the muscle contracts.

## Study TIPS

As you study the organizational structure of a muscle, think of its structure as a tube within a tube, within a tube, within a tube. In other words, the entire muscle is a tube. Within this tube are numerous "tubes" (fasciculi). Within the fasciculi are "tubes" (muscle fibers). As you examine the muscle fibers, you will note that they too are divided into "tubes" (myofibrils). Finally, the myofibrils are divided into even smaller "tubes" (myofilaments).

# C. NAMING SKELETAL MUSCLES

Muscle **nomenclature** is the system of naming skeletal muscle organs. The names of muscles, as with many organs, are in Latin rather than in English. By using only a handful of descriptive Latin words, commonly in combinations of two or three, anatomists have named all the muscles of the human body. Review these terms used to name muscles because they will come in handy when you learn specific muscle names in the next exercise.

☐ 1 Muscles can be named for their overall shape. Review these terms related to muscle shape:

| deltoid | shaped like **delta** (Δ) |
|---|---|
| orbicularis | circular |
| platy | flattened; platelike |
| quadratus | square |
| rhomboideus | diamond-shaped |
| trapezius | trapezoidal |
| triangularis | triangular |

☐ 2 Some muscles are named for their points of attachment (*origin* and *insertion*). For example, the *sternocleidomastoid* muscle has attachments on the sternum, clavicle, and mastoid process of the temporal bone.

☐ 3 Muscles can be named according to relative size:

| brevis | short |
|---|---|
| longus | long |
| magnus | large |
| maximus | largest |
| medius | moderately sized |
| minimus | small |

☐ 4 The direction of fibers visible in a muscle can be a basis for its name, using these terms:

| oblique | diagonal to the body's midline |
|---|---|
| rectus | parallel to the midline |
| sphincter | circling an opening |
| transversus | at a right angle to the midline |

☐ 5 Some muscle names are derived from the action(s) produced:

| abductor | abducts a part |
|---|---|
| adductor | adducts a part |
| depressor | depresses a part |
| extensor | extends a part |
| flexor | flexes a part |
| levator | elevates a part |
| rotator | rotates a part |

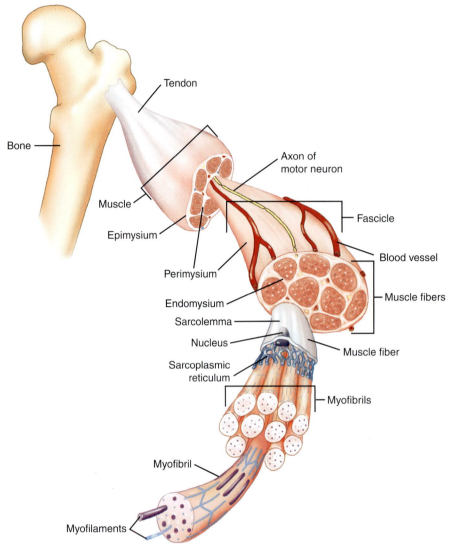

**FIGURE 17-2** Architecture of a muscle.

☐ 6 Some muscles are named for the region in which they are found. Some of these terms should be familiar to you:

| | |
|---|---|
| brachialis | arm |
| frontalis | frontal (bone) |
| femoris | femur |
| gluteus | posterior of hip/thigh |
| oculi | eye |
| radialis | radius |
| ulnaris | ulna |

✏ *Coloring Exercises:* **Skeletal Muscle**

Use colored pens or pencils to shade in both the figure and the labels. Each red numeral in the figure corresponds to a matching red numeral following the appropriate label.

TENDON₁
EPIMYSIUM₂
PERIMYSIUM₃
ENDOMYSIUM₄
MYOFIBRIL₅
THIN FILAMENT₆
THICK FILAMENT₇
SARCOMERE₈
A BAND₉
I BAND₁₀

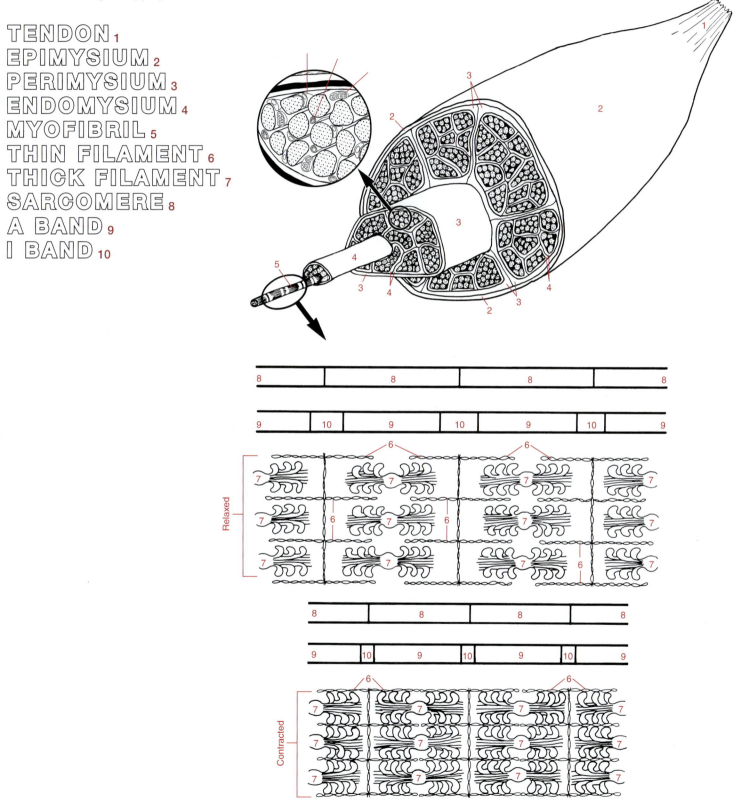

**FIGURE 17-3** Structure of skeletal muscle.

Name: _____ Date: _____ Section: _____

## LAB REPORT 17
# Organization of the Muscular System

## Fill-in

1. _____

2. _____

3. _____

4. _____

5. _____

6. _____

7. _____

8. _____

9. _____

10. _____

11. _____

12. _____

13. _____

14. _____

## Put in Order

1. _____

2. _____

3. _____

4. _____

## Fill-in (complete each item with the appropriate term)

1. The _?_ is the cell membrane of a muscle fiber.
2. A _?_ is a bundle of parallel myofilaments within a muscle fiber.
3. The _?_ is similar to the ER but collects and stores calcium ions in a resting muscle cell.
4. Two types of myofilaments in a skeletal muscle cell are _?_ filaments and thick filaments.
5. A repeating pattern, or unit, of myofibrils within the myofilament is the _?_.
6. A tube formed by the inward extension of the cell membrane is called a(n) _?_.
7. An entire muscle is covered with a fibrous sheath called the _?_.
8. A muscle is anchored by a fibrous band called a _?_, whereas it pulls at its insertion.
9. Each fascicle, or fasciculus, is wrapped with a fibrous sheath called a(n) _?_.
10. A broad, flat version of a tendon is more accurately called a(n) _?_.
11. Each individual muscle cell has a fibrous sheath called a(n) _?_ wrapped around its sarcolemma.
12. A muscle that flexes a joint may have the term _?_ in its name.
13. A muscle around an opening may be named by the term _?_.
14. A muscle associated with the femur may have _?_ in its name.

## Put in Order (rearrange these structures in anatomical order from superficial to deep)

sarcolemma
perimysium
endomysium
epimysium

**Identify**

1. _____
2. _____
3. _____
4. _____
5. _____
6. _____
7. _____
8. _____
9. _____
10. _____

**Identify** (determine the muscle naming term defined in each item)

1. Associated with the frontal bone
2. Adducts a part
3. Long
4. Encircles an opening or tube
5. Short
6. Flattened
7. Parallel to the midline of the body
8. Rotates a part
9. Associated with the ulna
10. Extends a part

# Skeletal Muscle

About 36% of the mass in female bodies and about 42% of the mass in male bodies is composed of skeletal muscle. In this exercise, some of the skeletal muscles of the human body are presented for study.

## BEFORE YOU BEGIN

- ☐ Read the appropriate chapter in your textbook.
- ☐ Set your learning goals. When you finish this exercise, you should be able to:
  - ☐ identify the following on a model and in figures:
    - ☐ muscles of the head and neck
    - ☐ muscles of the trunk
    - ☐ muscles of the upper extremity
    - ☐ muscles of the lower extremity
    - ☐ muscles of the pelvic floor
  - ☐ name the origin and insertion of each major muscle of the body
  - ☐ demonstrate the action of each muscle studied
- ☐ Prepare your materials:
  - ☐ models and charts of human musculature
  - ☐ colored pencils or pens
  - ☐ demonstration pointer
- ☐ Read the directions and safety tips for this exercise *carefully* before starting any procedure.

> **Hint** In this exercise, the muscles are presented by their location (e.g., head and trunk). Another common approach is to group muscles by the part they move (e.g., muscles that move the head or muscles that move the arm). Yet another manner of grouping muscles involves their action (e.g., flexors, extensors, and adductors). You may want to keep these groupings in mind as you explore the muscular system.
>
> Individual muscles can be located on a model of human musculature by locating them in the diagrams on the following pages, then checking their origins, insertions, and actions as listed on the tables for this exercise. Another method—more challenging, but more informative—is to try to identify them by their attachment points and actions *before* identifying them in a diagram. •

## A. IDENTIFYING MUSCLES

Using models and charts of the human musculature, locate the skeletal muscles outlined in the succeeding pages. As you do, note the size and shape of each muscle.

☐ 1 Identify these muscles of the *head and neck* **(Figure 18-1)**:
  - ☐ Occipitofrontalis (epicranius)
  - ☐ Orbicularis oculi
  - ☐ Orbicularis oris
  - ☐ Buccinator
  - ☐ Zygomaticus (two muscles: major and minor)
  - ☐ Levator labii superioris (two muscles: major and minor)
  - ☐ Depressor anguli oris
  - ☐ Temporalis
  - ☐ Masseter
  - ☐ Pterygoids (two muscles: internal and external)
  - ☐ Sternocleidomastoid
  - ☐ Trapezius

☐ 2 Identify these muscles of the *trunk* **(Figure 18-2)**:
  - ☐ Erector spinae (divides into three muscle groups)
  - ☐ Deep back muscles (several muscles)
  - ☐ External intercostals
  - ☐ Internal intercostals
  - ☐ Rectus abdominis
  - ☐ External abdominal oblique
  - ☐ Internal abdominal oblique
  - ☐ Transversus abdominis
  - ☐ Trapezius
  - ☐ Levator scapulae
  - ☐ Rhomboideus (two muscles: major and minor)
  - ☐ Serratus anterior
  - ☐ Pectoralis minor
  - ☐ Pectoralis major
  - ☐ Teres major
  - ☐ Latissimus dorsi
  - ☐ Infraspinatus
  - ☐ Supraspinatus
  - ☐ Subscapularis
  - ☐ Teres minor
  - ☐ Deltoid

### *Study* TIPS

Because muscle names have Latin origins, you should familiarize yourself with them. Your textbook has the word origins of key terms in each chapter. If you take the time to familiarize yourself with these, the names of the muscles will become easier to understand. Refer to the *A&P Survival Guide*.

> **Hint** Notice that many of the muscles listed for the trunk include shoulder (upper trunk) muscles that insert on the arm. •

## ✎ *Coloring Exercises:* Muscles of the Head and Neck

Use colored pens or pencils to shade in both the figure and the labels. Each red numeral in the figure corresponds to a matching red numeral following the appropriate label in **Table 18-1**.

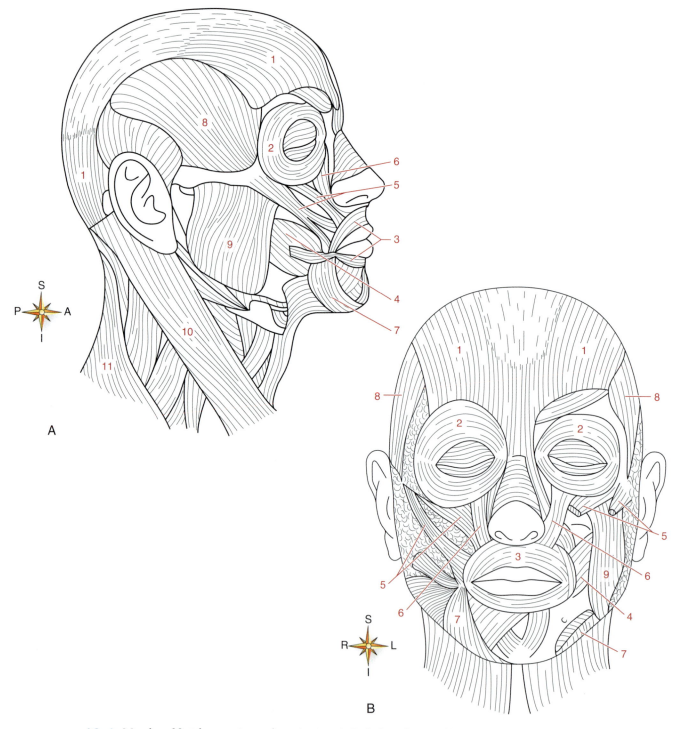

**FIGURE 18-1** Muscles of facial expression and mastication. **A,** Right lateral view of head and neck. **B,** Anterior view of head.

**TABLE 18-1**   **Muscles of the Head and Neck**

| MUSCLE | ORIGIN | INSERTION | ACTION |
| --- | --- | --- | --- |
| OCCIPITOFRONTALIS 1 | Occipital | Skin of eyebrow, nose | Elevates brows; moves scalp |
| ORBICULARIS OCULI 2 | Maxilla, frontal | Encircles eye, near origin | Closes eye |
| ORBICULARIS ORIS 3 | Maxilla, mandible | Lips | Closes lips |
| BUCCINATOR 4 | Maxilla, mandible | Angle of mouth | Compresses cheeks |
| ZYGOMATICUS 5 (TWO) | Zygomatic bone | Angle of mouth, upper lip | Elevates angle of mouth, upper lip |
| LEVATOR LABII SUPERIORIS 6 | Maxilla | Upper lip, nose | Elevates upper lip, nose |
| DEPRESSOR ANGULI ORIS 7 | Mandible | Lower lip near angle | Depresses angle of mouth |
| TEMPORALIS 8 | Temporal aspect of skull | Mandible | Closes jaw |
| MASSETER 9 | Zygomatic arch | Mandible | Closes jaw |
| PTERYGOIDS (TWO) | Inferior aspect of skull | Mandible | Medial closes jaw; lateral opens jaw |
| STERNOCLEIDOMASTOID 10 | Sternum, clavicle | Mastoid process | Rotates, flexes neck |
| TRAPEZIUS 11 | Skull, upper vertebral column | Scapula | Extends head, neck |

Use terms in OUTLINE type as coloring labels. Terms in **SOLID** type do not appear in the coloring plate.

*Coloring Exercises:* **Muscles of the Trunk**

Use colored pens or pencils to shade in both the figure and the labels. Each red numeral in the figure corresponds to a matching red numeral following the appropriate label in **Table 18-2**.

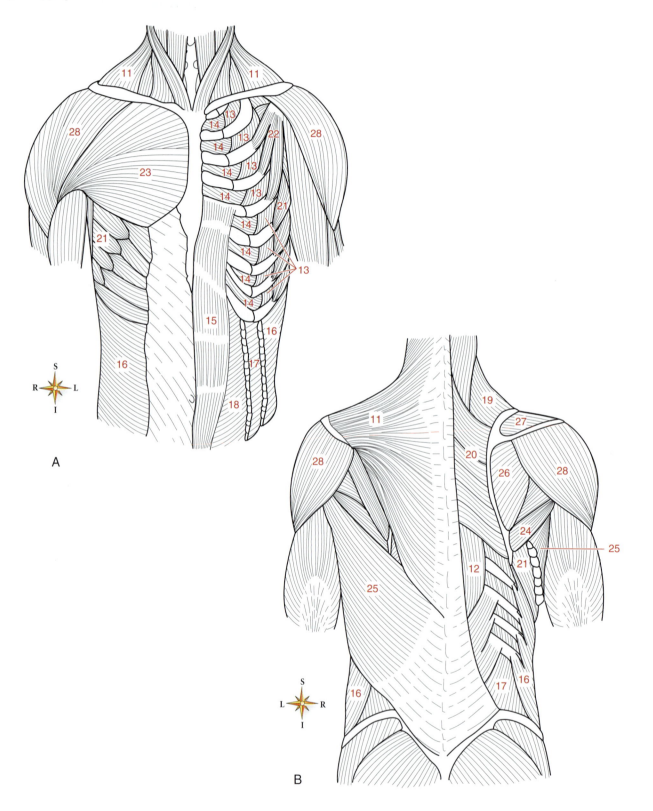

**FIGURE 18-2** Muscles of the trunk and abdominal wall. **A,** Anterior view of trunk. **B,** Posterior view of trunk.

**TABLE 18-2**   Muscles of the Trunk

| MUSCLE | ORIGIN | INSERTION | ACTION |
|---|---|---|---|
| ERECTOR SPINAE 12 (DIVIDES INTO THREE GROUPS) | Vertebrae, pelvis | Superior vertebrae, ribs | Holds body upright |
| DEEP BACK MUSCLES | Vertebrae | Vertebrae | Flexes or extends trunk |
| EXTERNAL INTERCOSTALS 13 | Ribs | Edge of next rib (inferiorly) | Expands thorax |
| INTERNAL INTERCOSTALS 14 | Ribs | Edge of next rib (superiorly) | Compresses thorax |
| RECTUS ABDOMINIS 15 | Pubis | Inferior thoracic cage | Flexes waist |
| EXTERNAL ABDOMINAL OBLIQUE 16 | Inferior thoracic cage | Midline of abdomen | Compresses abdomen |
| INTERNAL ABDOMINAL OBLIQUE 17 | Pelvis | Midline of abdomen | Compresses abdomen |
| TRANSVERSUS ABDOMINIS 18 | Vertebrae, pelvis, ribs | Midline of abdomen | Compresses abdomen |
| TRAPEZIUS 11 | Skull, upper vertebral column | Scapula | Extends head, neck; elevates, depresses, or rotates scapula |
| LEVATOR SCAPULAE 19 | Vertebrae | Scapula | Elevates scapula |
| RHOMBOIDEUS 20 (TWO) | Vertebrae | Scapula | Retracts scapula |
| SERRATUS ANTERIOR 21 | Ribs | Scapula | Protracts scapula |
| PECTORALIS MINOR 22 | Ribs | Scapula | Depresses scapula |
| PECTORALIS MAJOR 23 | Ribs, clavicle | Humerus | Adducts, flexes arm |
| TERES MAJOR 24 | Scapula | Humerus | Extends, adducts, rotates arm |
| LATISSIMUS DORSI 25 | Vertebrae | Humerus | Extends, adducts arm |
| INFRASPINATUS 26 | Scapula | Humerus | Extends, rotates arm |
| SUPRASPINATUS 27 | Scapula | Humerus | Abducts arm |
| SUBSCAPULARIS | Scapula | Humerus | Extends, rotates arm |
| TERES MINOR | Scapula | Humerus | Adducts, rotates arm |
| DELTOID 28 | Scapula, clavicle | Humerus | Abducts arm |

Use terms in OUTLINE type as coloring labels. Terms in **SOLID** type do not appear in the coloring plate.

✎ *Coloring Exercises:* **Muscles of the Upper Extremity**

Use colored pens or pencils to shade in both the figure and the labels. Each red numeral in the figure corresponds to a matching red numeral following the appropriate label in **Table 18-3**.

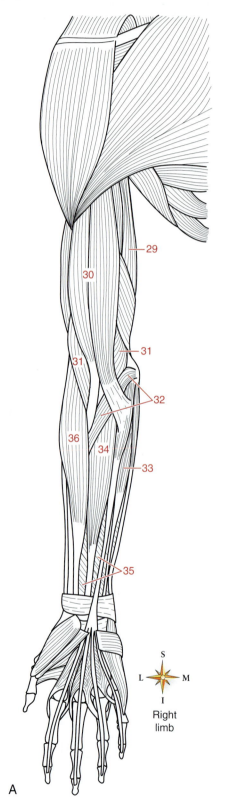

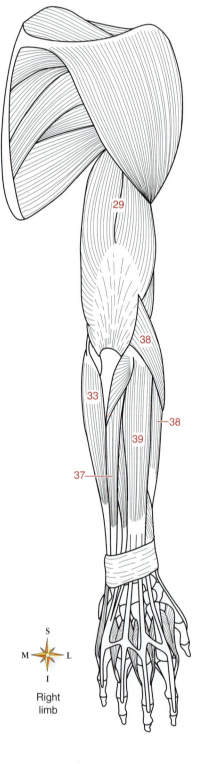

A

B

**FIGURE 18-3** Muscles of the arm. **A,** Anterior view of right upper extremity. **B,** Posterior view of right upper extremity.

**TABLE 18-3**  **Muscles of the Upper Extremity**

| MUSCLE | ORIGIN | INSERTION | ACTION |
|---|---|---|---|
| TRICEPS BRACHII 29 | Scapula | Ulna | Extends forearm |
| BICEPS BRACHII 30 | Humerus, scapula | Radius | Flexes, supinates forearm |
| BRACHIALIS 31 | Humerus | Ulna | Flexes forearm |
| PRONATORS 32 (TWO) | Ulna, humerus | Radius | Pronates forearm |
| FLEXOR CARPI ULNARIS 33 | Humerus (medial epicondyle) | Carpal bone | Flexes, abducts wrist |
| FLEXOR CARPI RADIALIS 34 | Humerus (medial epicondyle) | Metacarpal bones | Flexes, abducts wrist |
| FLEXOR DIGITORUM 35 (TWO) | Humerus (medial epicondyle, ulna, radius) | Phalanges | Flexes fingers |
| BRACHIORADIALIS 36 | Humerus | Radius (distal) | Flexes forearm |
| SUPINATOR | Ulna | Radius | Supinates forearm |
| EXTENSOR CARPI ULNARIS 37 | Humerus (lateral epicondyle) | Metacarpal bones | Extends, abducts wrist |
| EXTENSOR CARPI RADIALIS 38 (TWO) | Humerus (lateral epicondyle) | Metacarpal bones | Extends, abducts wrist |
| EXTENSOR DIGITORUM 39 | Humerus (lateral epicondyle) | Phalanges | Extends fingers |

Use terms in OUTLINE type as coloring labels. Terms in **SOLID** type do not appear in the coloring plate.

# Coloring Exercises: Muscles of the Pelvic Floor

Use colored pens or pencils to shade in both the figure and the labels. Each red numeral in the figure corresponds to a matching red numeral following the appropriate label in **Table 18-5**.

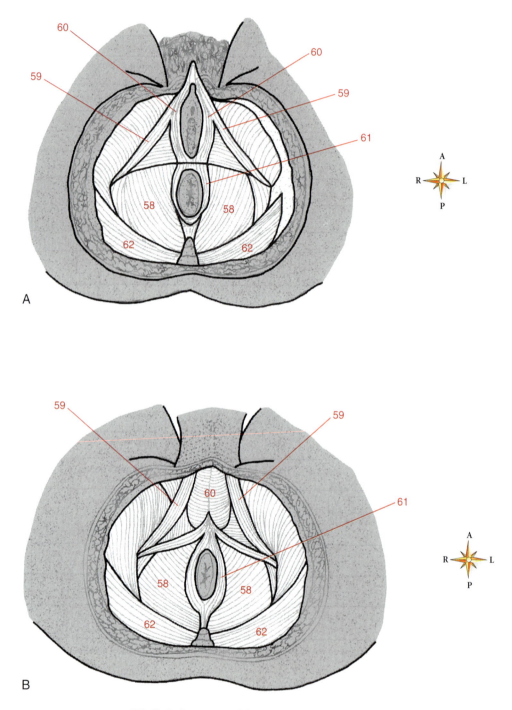

**FIGURE 18-5**  Inferior view of the pelvic muscles. **A**, Female. **B**, Male.

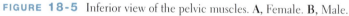

**Hint** By now you have noticed that several of the muscles illustrated in the coloring exercises do not have labels. Some of them are muscles that you were asked to identify but are not labeled for coloring because they are too far away from the outline labels. Go back to the figures and try to identify the unlabeled muscles. Use your textbook and anatomy reference books in your school library to help you.  •

**TABLE 18-5**    Muscles of the Pelvic Floor

| MUSCLE | ORIGIN | INSERTION | ACTION |
|---|---|---|---|
| LEVATOR ANI 58 | Pubis, ischium | Sacrum, coccyx | Elevates anus |
| ISCHIOCAVERNOSUS 59 | Ischium | Clitoris or penis | Compresses base of clitoris or penis |
| BULBOSPONGIOSUS 60 | MALE: bulb of penis FEMALE: central tendon of perineum | MALE: central tendon of perineum FEMALE: base of clitoris | MALE: constricts urethra; erects penis FEMALE: erects clitoris |
| SPHINCTER EXTERNUS ANI 61 | Coccyx | Central tendon (median raphe) | Closes anal canal |
| GLUTEUS MAXIMUS 62 | Hip | Femur | Extends thigh |

Use terms in OUTLINE type as coloring labels.

☐  3  Identify these muscles of the *upper extremity* (Figure 18-3):
☐  Triceps brachii
☐  Biceps brachii
☐  Brachialis
☐  Pronators (two muscles: teres and quadratus)
☐  Flexor carpi ulnaris
☐  Flexor carpi radialis
☐  Flexor digitorum (two muscles: longus and brevis)
☐  Brachioradialis
☐  Supinator
☐  Extensor carpi ulnaris
☐  Extensor carpi radialis (two muscles: profundus and superficialis)
☐  Extensor digitorum
☐  4  Identify these muscles of the *lower extremity* (Figure 18-4):
☐  Iliopsoas
☐  Tensor fasciae latae
☐  Gluteal group: gluteus maximus, gluteus medius, gluteus minimus
☐  Quadriceps femoris group: rectus femoris, vastus lateralis, vastus medialis, vastus intermedius
☐  Sartorius
☐  Hamstring group: biceps femoris, semimembranosus, semitendinosus
☐  Adductor group: adductor longus, gracilis
☐  Tibialis anterior
☐  Extensor digitorum longus

☐  Gastrocnemius
☐  Soleus
☐  Fibularis (three muscles: longus, brevis, and tertius)
☐  5  Identify these muscles of the *pelvic floor* (Figure 18-5):
☐  Levator ani
☐  Ischiocavernosus
☐  Bulbospongiosus
☐  Sphincter externus ani
☐  Gluteus maximus

## B. DEMONSTRATING MUSCLE ACTION

For each of the muscles found in Activity A of this exercise, demonstrate its action with your own muscle (if possible). As you contract the muscle, palpate it and note its size and location.

**safety first**
Be careful to avoid injuring yourself and others as you perform each muscle action.  •

**Hint** ▶ For a more detailed description of the origin, insertion, action, and innervation of some of these muscles, please refer to your textbook.  •

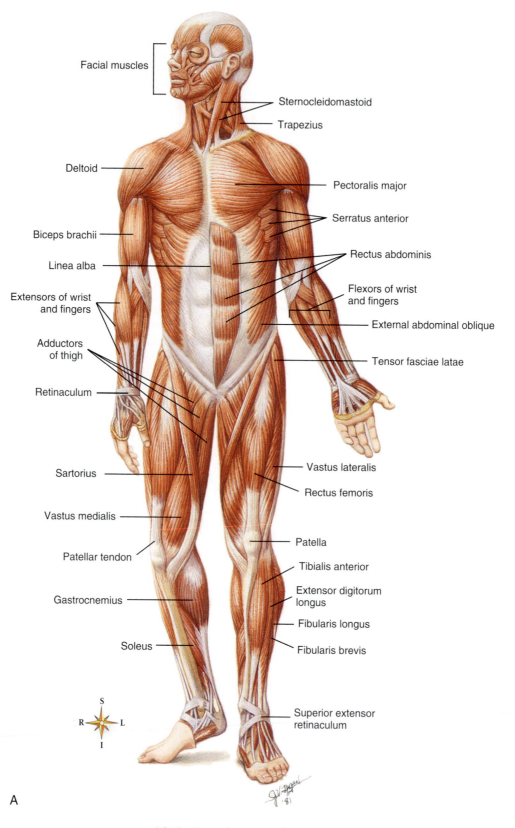

Facial muscles
Sternocleidomastoid
Trapezius
Deltoid
Pectoralis major
Serratus anterior
Biceps brachii
Linea alba
Rectus abdominis
Extensors of wrist and fingers
Flexors of wrist and fingers
Adductors of thigh
External abdominal oblique
Retinaculum
Tensor fasciae latae
Sartorius
Vastus lateralis
Rectus femoris
Vastus medialis
Patellar tendon
Patella
Tibialis anterior
Extensor digitorum longus
Gastrocnemius
Fibularis longus
Fibularis brevis
Soleus
Superior extensor retinaculum

A

**FIGURE 18-6** General overview of the body. **A,** Anterior view.

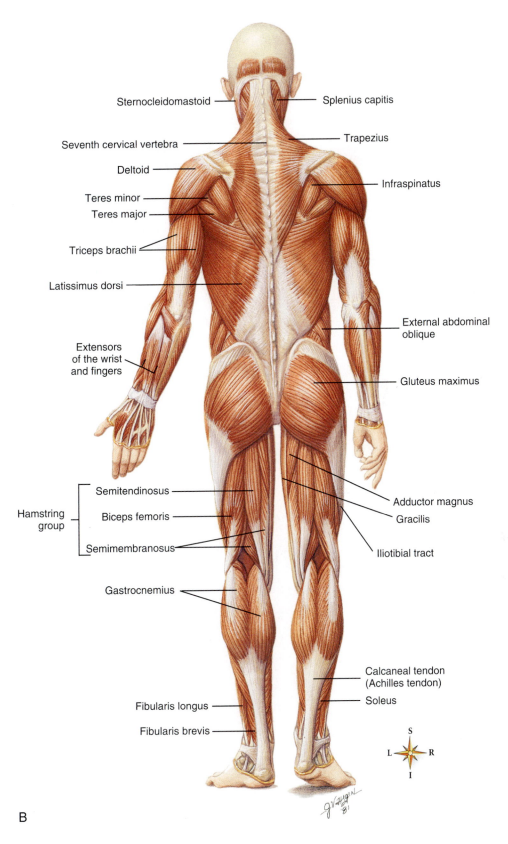

Sternocleidomastoid

Splenius capitis

Seventh cervical vertebra

Trapezius

Deltoid

Infraspinatus

Teres minor

Teres major

Triceps brachii

Latissimus dorsi

External abdominal oblique

Extensors of the wrist and fingers

Gluteus maximus

Semitendinosus

Hamstring group

Biceps femoris

Adductor magnus

Gracilis

Semimembranosus

Iliotibial tract

Gastrocnemius

Calcaneal tendon (Achilles tendon)

Soleus

Fibularis longus

Fibularis brevis

**FIGURE 18-6, cont'd B,** Posterior view.

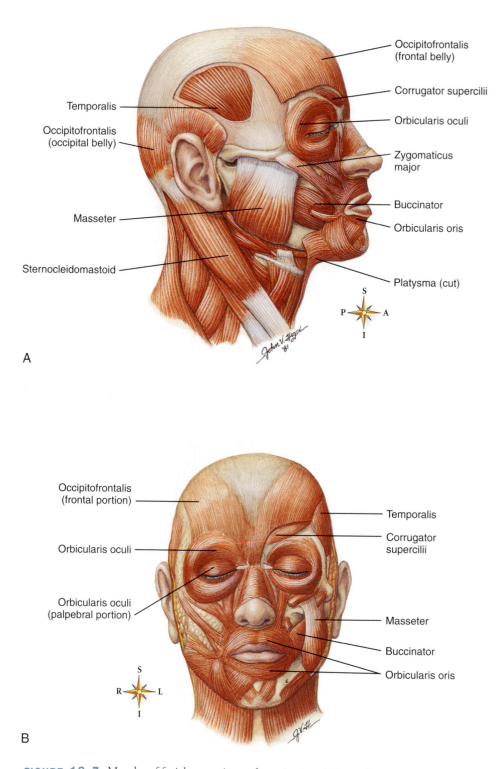

Occipitofrontalis
(frontal belly)

Corrugator supercilii

Orbicularis oculi

Zygomaticus
major

Buccinator

Orbicularis oris

Platysma (cut)

Temporalis

Occipitofrontalis
(occipital belly)

Masseter

Sternocleidomastoid

A

Occipitofrontalis
(frontal portion)

Orbicularis oculi

Orbicularis oculi
(palpebral portion)

Temporalis

Corrugator
supercilii

Masseter

Buccinator

Orbicularis oris

B

**FIGURE 18-7** Muscles of facial expression and mastication. **A,** Lateral view. **B,** Anterior view.

Name: _____ Date: _____ Section: _____

# Skeletal Muscle Identification

**Table Completion** (list each muscle name or muscle combination/group name only once in the table below; also, some boxes will not have an answer)

Adductor group  
Latissimus dorsi  
Biceps brachii  
Pectoralis major  
Deltoid  
Pectoralis major/latissimus dorsi combination  
Erector spinae  
Quadriceps femoris group  

Gastrocnemius  
Rectus abdominis  
Gluteus maximus  
Tensor fasciae latae  
Hamstring group  
Tibialis anterior  
Iliopsoas  
Triceps brachii  

| PART MOVED | FLEXOR | EXTENSOR | ABDUCTOR | ADDUCTOR |
|---|---|---|---|---|
| Arm |  |  |  |  |
| Forearm |  |  |  |  |
| Thigh |  |  |  |  |
| Leg |  |  |  |  |
| Foot |  |  |  |  |
| Trunk |  |  |  |  |

## Fill-in

1. _____
2. _____
3. _____
4. _____
5. _____
6. _____
7. _____
8. _____
9. _____
10. _____
11. _____
12. _____
13. _____
14. _____
15. _____
16. _____
17. _____
18. _____
19. _____
20. _____
21. _____
22. _____
23. _____
24. _____
25. _____

## Fill-in (complete each item with the correct term)

1. The occipitofrontalis, or epicranius, originates on the _?_ bone.
2. The masseter, temporalis, and pterygoids all insert on the _?_.
3. The sternocleidomastoid inserts on the _?_.
4. The temporalis closes the _?_.
5. The _?_ closes the eye.
6. The pectoralis _?_ inserts on the humerus.
7. The deltoid inserts on the _?_.
8. The _?_ intercostals help accomplish forced expiration.
9. The erector _?_ help maintain posture.
10. The pectoralis major is _?_ to the pectoralis minor.
11. The latissimus dorsi is on the _?_ side of the trunk.
12. The _?_ intercostals contract when you take a deep breath.
13. The rippling effect seen on the lower midline of the abdomens of some athletes is caused by hypertrophy of the _?_ muscle.
14. The origin of the brachialis is _?_ to its insertion.
15. In grasping a baseball tightly in the hand, one would likely use the _?_ muscle.
16. Both the biceps brachii and the _?_ flex the forearm.
17. The extensor digitorum inserts on the _?_.
18. The adductor longus and the _?_ are both part of the adductor group.
19. The hamstring muscles all _?_ the leg.
20. The muscle that pulls the leg so that you can cross it over the other leg while sitting is called the _?_ muscle.
21. Both the gastrocnemius and the _?_ plantar flex the foot.
22. Muscles of the quadriceps group all _?_ the leg.
23. In the male, the _?_ pulls the penis erect but also can constrict the urethra.
24. The large muscle that extends the thigh is the _?_.
25. The levator ani raises the _?_.

# LAB EXERCISE 19
# Dissection: Skeletal Muscles

In the previous exercise, you identified the components of the human musculature on models and/or charts of the human muscular system. In this exercise, you are challenged to find some of the major muscles in a preserved mammalian specimen. While the musculature of these animals may differ somewhat from human musculature, a dissection exercise stimulates an appreciation for the highly variable, three-dimensional nature of muscular anatomy that no other type of learning activity can.

Activity A provides directions for studying the musculature of the cat. Activity B is an alternate activity, providing directions for studying the musculature of the fetal pig.

## BEFORE YOU BEGIN

☐ Read the appropriate chapter in your textbook.
☐ Set your learning goals. When you finish this exercise, you should be able to:
    ☐ dissect the musculature of a preserved cat or fetal pig
    ☐ identify the following in a dissected mammalian specimen:
        ☐ muscles of the head and neck
        ☐ muscles of the trunk
        ☐ muscles of the upper extremity
        ☐ muscles of the lower extremity
        ☐ muscles of the pelvic floor
☐ Prepare your materials:
    ☐ preserved (plain or injected) cat or fetal pig
    ☐ dissection tools and trays
    ☐ protective gear: gloves, safety eyewear, and apron
    ☐ mounted skeleton of the cat or fetal pig
    ☐ storage container (if specimen is to be reused)
☐ Read the directions and safety tips for this exercise *carefully* before starting any procedure.

> ⚠ **safety first**
> Observe the usual precautions when working with a preserved specimen. Heed the safety advice accompanying the preservatives used with your specimen. Use protective gloves while handling your specimen. Use safety goggles to avoid injury during dissections. Avoid injury with dissection tools. Dispose of or store your specimen as instructed. •

## *Study* TIPS

As you complete this dissection, refer to an illustration or atlas of the human body. Use the illustration or atlas to locate and compare the human musculature with your preserved specimen.

## A. CAT MUSCULATURE

If your cat was not skinned in Lab Exercise 10—The Skin, it must be done before beginning this activity. Refer to Lab Exercise 10 for directions regarding the removal of the skin (see **Figure 10-5**).

> **Hint** ▶ The muscles are held in place by various connective tissues. The best way to dissect muscles is to first scrape or cut away any fascia that covers the muscle of interest, *taking care not to cut into the muscle itself*. Next, carefully slip a blunt probe under the muscle and slide it back and forth to break the loose fibers holding the muscle in place. It is best not to cut any of the muscles unless you absolutely must to see an important underlying muscle. When you do cut a muscle, cut it in a way that will permit you to later fold it back into its normal position. Remember, you will probably be studying these muscles over and over again as you prepare for your practical examination on cat musculature. A bag full of separate muscle organs will not help you identify them by location in another specimen. •

☐ 1 Identify the muscles of the *head and neck* **(Table 19-1)**. Before you begin, you must place the cat on its back and remove the skin and underlying connective tissue up to the chin **(Figure 19-2)**.
    ☐ Mylohyoid
    ☐ Masseter
    ☐ Digastric
    ☐ Sternohyoid
    ☐ Sternomastoid

> **Hint** ▶ Refer to **Figures 19-2 to 19-6** for color photographs of dissected cat musculature. •

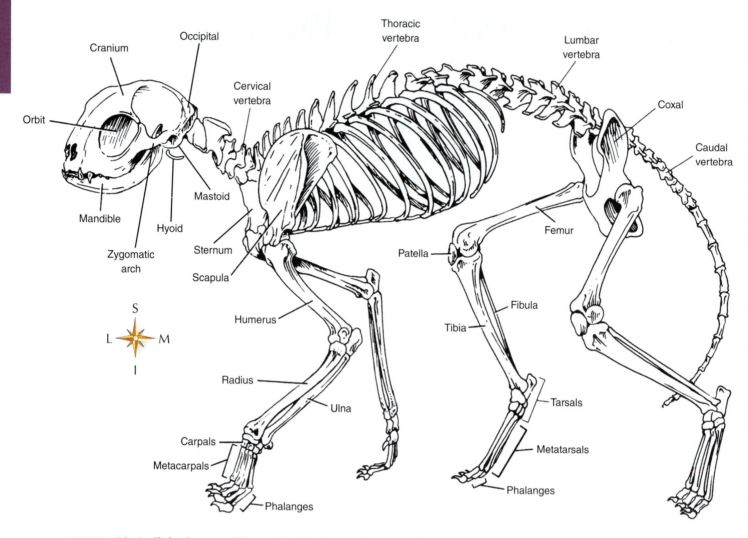

**FIGURE 19-1** Skeletal system of the cat. Use this figure to locate muscle origins and insertions listed in the tables in this exercise.

**TABLE 19-1**   Muscles of the Head and Neck of the Cat

| MUSCLE | ORIGIN | INSERTION | ACTION |
| --- | --- | --- | --- |
| Mylohyoid | Mandible (body) | Hyoid | Depresses mandible; elevates floor of mouth |
| Masseter | Zygomatic | Mandible | Elevates, protracts mandible |
| Digastric | Occipital | Mandible | Depresses, retracts mandible |
| Sternohyoid | Sternum | Hyoid | Depresses hyoid |
| Sternomastoid | Sternum | Temporal (mastoid) | Flexes, rotates head |

☐ 2 With the cat still on its back, identify the superficial muscles of the *trunk and shoulder* (ventral aspect) **(Table 19-2)**.
    ☐ Clavodeltoid (clavobrachialis)
    ☐ Pectoantebrachialis
    ☐ Pectoralis major
    ☐ Pectoralis minor
    ☐ Xiphihumeralis
    ☐ Serratus ventralis
    ☐ External oblique

☐ 3 Make careful incisions near the midline to expose the following deeper muscles of the *trunk* (ventral aspect):
    ☐ Internal oblique
    ☐ Transversus abdominis
    ☐ Rectus abdominis

☐ 4 Turn the cat onto its ventral side and identify the following superficial muscles of the dorsal aspect of the *trunk and shoulder* **(Figure 19-4)**:
    ☐ Clavotrapezius
    ☐ Levator scapulae ventralis
    ☐ Acromiotrapezius
    ☐ Spinotrapezius
    ☐ Clavodeltoid
    ☐ Acromiodeltoid
    ☐ Spinodeltoid
    ☐ Latissimus dorsi

☐ 5 With your scissors, cut the three trapezius muscles (see **Figure 19-4, B**) to expose the deeper muscles of the *trunk and shoulder*. Identify these muscles:
    ☐ Supraspinatus
    ☐ Infraspinatus
    ☐ Teres major
    ☐ Rhomboideus major
    ☐ Rhomboideus minor
    ☐ Rhomboideus capitis
    ☐ Splenius

☐ 6 Turn the cat so that it is again lying on its back and identify these muscles of the *forelimb* from the medial aspect **(Figure 19-5, A, B,** and **Table 19-3)**:
    ☐ Triceps brachii
    ☐ Epitrochlearis
    ☐ Biceps brachii

☐ 7 Place the cat so that it is lying on its ventral surface and identify these muscles of the *forelimb* from the lateral aspect (see **Figure 19-5, C-E,** and **Table 19-3**):
    ☐ Brachialis
    ☐ Triceps brachii

> **Hint** You will have to cut some of the superficial muscles of the upper limb to see the deeper muscles. See **Figure 19-5** for examples. •

☐ 8 Place the cat on its back and identify these superficial and deep muscles of the *hindlimb* from the medial aspect (**Figure 19-6** and **Table 19-4**):
    ☐ Sartorius
    ☐ Gracilis
    ☐ Rectus femoris
    ☐ Vastus medialis
    ☐ Adductor longus
    ☐ Semimembranosus
    ☐ Semitendinosus
    ☐ Tensor fasciae latae
    ☐ Flexor digitorum longus
    ☐ Gastrocnemius
    ☐ Soleus
    ☐ Tibialis anterior

☐ 9 Place the cat on its ventral surface and identify these muscles of the *hindlimb* from the lateral aspect (see **Figure 19-6, C**):
    ☐ Gluteus medius
    ☐ Gluteus maximus
    ☐ Caudofemoralis
    ☐ Tensor fasciae latae
    ☐ Biceps femoris
    ☐ Semimembranosus
    ☐ Semitendinosus
    ☐ Gastrocnemius
    ☐ Soleus
    ☐ Peroneus
    ☐ Tibialis anterior

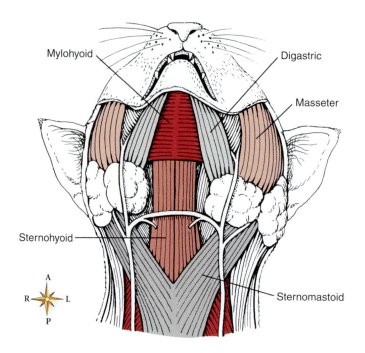

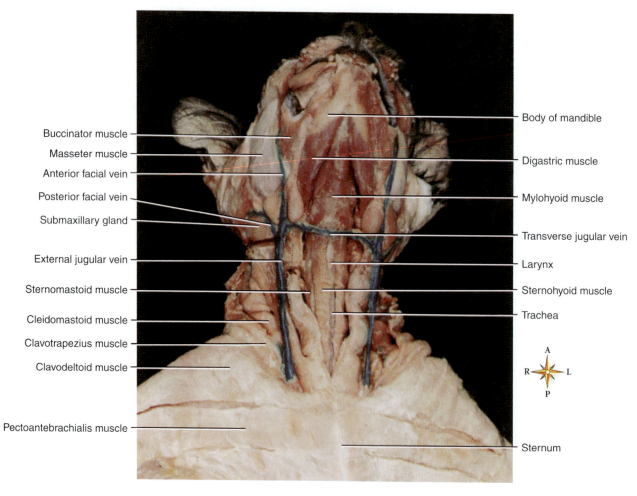

**FIGURE 19-2** Muscles of the head and neck of the cat, ventral view.

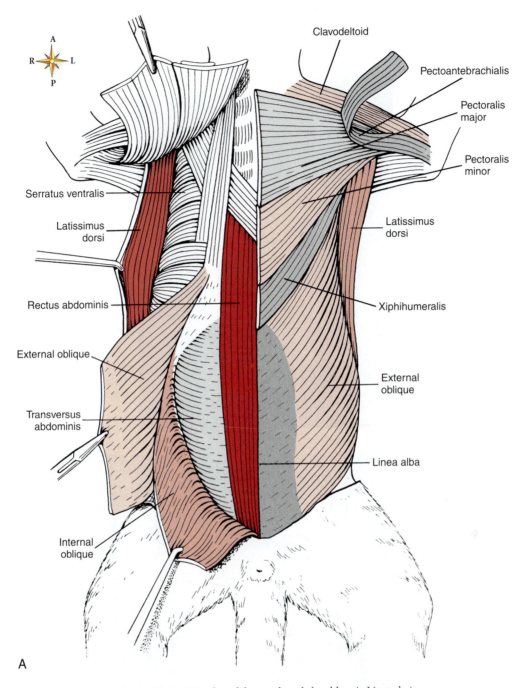

A

Clavodeltoid

Pectoantebrachialis

Pectoralis major

Pectoralis minor

Serratus ventralis

Latissimus dorsi

Latissimus dorsi

Rectus abdominis

Xiphihumeralis

External oblique

External oblique

Transversus abdominis

Linea alba

Internal oblique

**FIGURE 19-3** Muscles of the trunk and shoulder. **A,** Ventral view.                    *Continued*

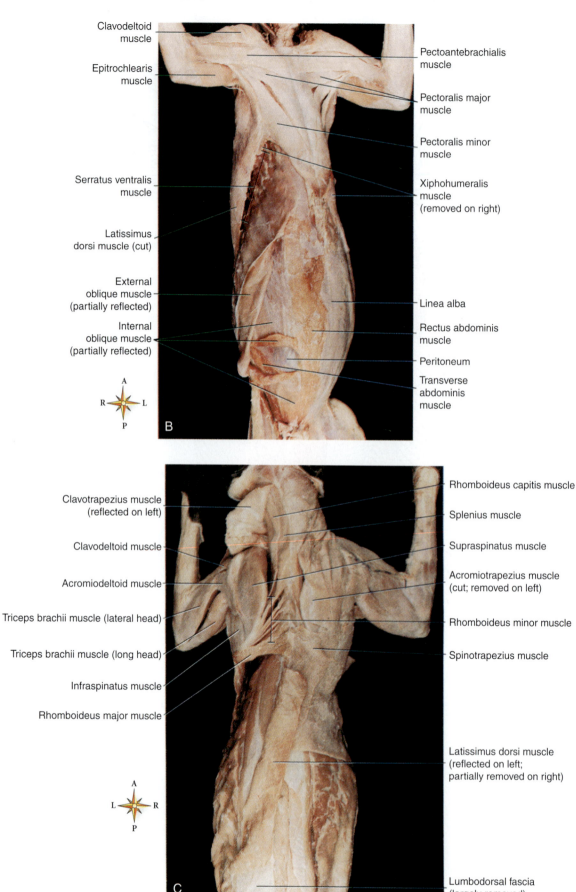

Clavodeltoid muscle

Epitrochlearis muscle

Serratus ventralis muscle

Latissimus dorsi muscle (cut)

External oblique muscle (partially reflected)

Internal oblique muscle (partially reflected)

A
R — L
P

B

Pectoantebrachialis muscle

Pectoralis major muscle

Pectoralis minor muscle

Xiphohumeralis muscle (removed on right)

Linea alba

Rectus abdominis muscle

Peritoneum

Transverse abdominis muscle

Clavotrapezius muscle (reflected on left)

Clavodeltoid muscle

Acromiodeltoid muscle

Triceps brachii muscle (lateral head)

Triceps brachii muscle (long head)

Infraspinatus muscle

Rhomboideus major muscle

A
L — R
P

C

Rhomboideus capitis muscle

Splenius muscle

Supraspinatus muscle

Acromiotrapezius muscle (cut; removed on left)

Rhomboideus minor muscle

Spinotrapezius muscle

Latissimus dorsi muscle (reflected on left; partially removed on right)

Lumbodorsal fascia (largely removed)

**FIGURE 19-3, cont'd** B, Ventral view. C, Dorsal view.

**T A B L E   1 9 - 2**   Muscles of the Trunk and Shoulder of the Cat

| MUSCLE | ORIGIN | INSERTION | ACTION |
|---|---|---|---|
| Clavodeltoid | Clavicle | Ulna (proximal end) | Extends humerus; flexes elbow; rotates head |
| Pectoantebrachialis | Sternum (anterior) | Ulnar fascia | Adducts humerus |
| Pectoralis major | Sternum, clavicle | Humerus | Adducts humerus |
| Pectoralis minor | Sternum | Humerus | Adducts humerus |
| Xiphihumeralis | Sternum (xiphoid) | Humerus | Adducts humerus |
| Serratus ventralis | Ribs 1-9 | Scapula (medial) | Depresses scapula |
| External oblique | Lumbodorsal fascia, posterior ribs | Linea alba | Compresses abdomen; flexes, rotates vertebral column |
| Internal oblique | Lumbodorsal fascia, pelvis | Linea alba | Compresses abdomen; flexes, rotates vertebral column |
| Transversus abdominis | Lumbar vertebra, costal cartilage | Linea alba | Compresses abdomen |
| Rectus abdominis | Coxal (pubis) | Sternum, costal cartilages | Compresses abdomen |
| Clavotrapezius | Back of skull, dorsal midline of neck | Clavicle | Moves head and clavicle toward each other |
| Levator scapulae ventralis | Occipital and vertebra C1 (atlas) | Scapula | Moves scapula anteriorly |
| Acromiotrapezius | Vertebrae C2 to T3 (spinous processes) | Scapula (spine) | Moves scapula toward midline of body |
| Spinotrapezius | Vertebrae C1 to C4 and T1 to T12 (spinous processes) | Scapula | Moves scapula toward midline of body and posteriorly |
| Acromiodeltoid | Scapula (acromion) | Humerus (proximal end) | Adducts, rotates humerus |
| Spinodeltoid | Scapula (spine) | Humerus | Elevates, rotates humerus |
| Latissimus dorsi | Lumbodorsal fascia and vertebrae T4 to L6 | Humerus | Adducts, rotates, extends humerus |
| Supraspinatus | Scapula (above spine) | Humerus (greater tubercle) | Extends humerus; moves scapula toward head |
| Infraspinatus | Scapula (below spine) | Humerus (greater tubercle) | Rotates humerus |
| Teres major | Scapula (lateral border) | Humerus (proximal end) | Rotates humerus |
| Rhomboideus major | Vertebrae (thoracic) | Scapula (inferior angle) | Pulls scapula toward vertebral column |
| Rhomboideus minor | Vertebrae (posterior cervical and thoracic) | Scapula (vertebral border) | Pulls scapula toward vertebral column |
| Rhomboideus capitis | Occipital | Scapula (vertebral border) | Pulls scapula forward and medially |
| Splenius | Vertebrae (T1 and T2) and fascia of neck | Occipital | Extends and turns head laterally |

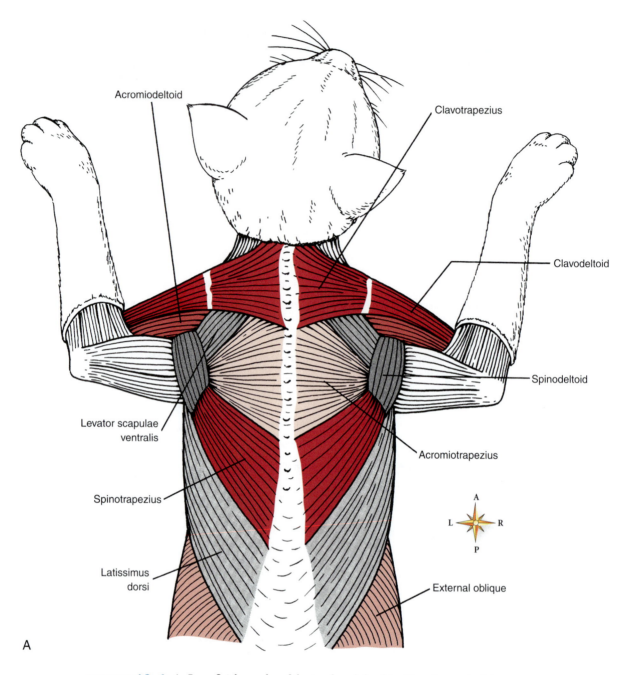

**FIGURE 19-4** **A,** Superficial muscles of the trunk and shoulder (dorsal aspect) of the cat.

Acromiodeltoid

Clavotrapezius

Clavodeltoid

Spinodeltoid

Levator scapulae ventralis

Acromiotrapezius

Spinotrapezius

Latissimus dorsi

External oblique

A

Lumbodorsal fasci

Gluteus medius musc

Sartorius musc

Tensor fasciae latae musc

Gluteus maximus musc

Caudofemoralis musc

Biceps femoris musc

Semitendinosus musc

Semimembranosus musc

Gastrocnemius musc

Soleus musc

Tibialis anterior musc

Peroneus musc

Achilles tend

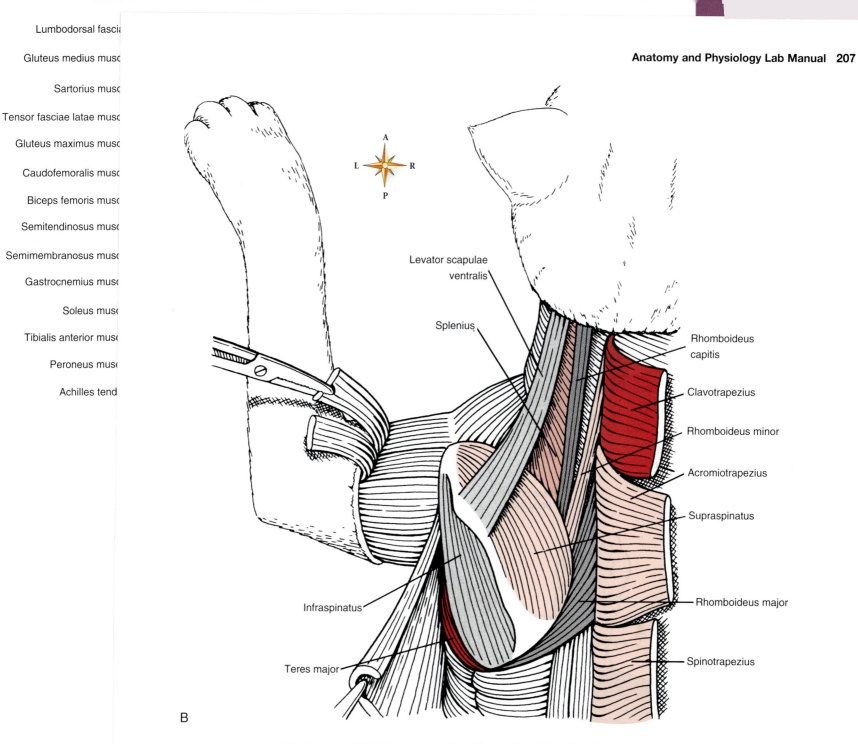

Levator scapulae ventralis

Splenius

Rhomboideus capitis

Clavotrapezius

Rhomboideus minor

Acromiotrapezius

Supraspinatus

Rhomboideus major

Spinotrapezius

Infraspinatus

Teres major

B

**FIGURE 19-4, cont'd** **B,** Deeper muscles of the trunk and shoulder (left dorsal aspect) of the cat.

Adductor

Pec

Adductor f

Gracil

Semimembra

Semitend

Biceps f

Gastrocnemius mus

# 20
# Skeletal Muscle Contractions

Now that you are familiar with the structure of the muscular system and its organs, it is time to take a closer look at the physiology of this system. The first activity demonstrates contraction at the cell level. The second and third activities challenge you to investigate concepts of muscle function at the organ level.

## BEFORE YOU BEGIN

- ☐ Read the appropriate chapter in your textbook.
- ☐ Set your learning goals. When you finish this exercise, you should be able to:
  - ☐ describe some factors that influence the contraction of single skeletal muscle fibers
  - ☐ interpret a myogram of a single twitch, determining its three phases
  - ☐ understand the nature of the treppe phenomenon and tetanus
  - ☐ demonstrate how muscle contraction can be studied in a laboratory setting
- ☐ Prepare your materials:
  - ☐ For Activity A:
    - ☐ glycerinated rabbit psoas muscle
    - ☐ watch glass or petri dish
    - ☐ microscopes (dissecting and compound light microscope)
    - ☐ fine glass probes
    - ☐ glass slides and coverslips
    - ☐ glycerin solution (50% in triple distilled water)
    - ☐ metric ruler (mm)
    - ☐ adenosine triphosphate (ATP) + salt solution
    - ☐ ATP solution
    - ☐ salt solution
  - ☐ For Activity B:
    - ☐ frog (live)
    - ☐ pithing needle
    - ☐ dissection pan or board
    - ☐ dissection tools
    - ☐ Ringer's solution (amphibian)
    - ☐ blunt glass probe or polished tube
    - ☐ string
    - ☐ kymograph or physiograph system
    - ☐ stimulator apparatus
  - ☐ For Activity C:
    - ☐ grip strength device
    - ☐ computer and access to a printer
    - ☐ electrode gel
    - ☐ flat plate electrode (and tape)
    - ☐ stimulator apparatus
    - ☐ pen (washable ink)
    - ☐ USB drive/thumb drive (if data are to be stored)
- ☐ Read the directions and safety tips for this exercise *carefully* before starting any procedure.

### *Study* TIPS

Before you begin this lab, review the physiology of muscle contraction. Understand what chemicals are necessary for muscle contraction. Refer to your textbook or the *A&P Survival Guide*.

## A. CONTRACTION OF SINGLE MUSCLE FIBER

In this activity, you will observe contractions in single skeletal muscle fibers. Your specimen is a special glycerinated preparation of a rabbit psoas muscle.

> ⓘ **safety first**
> Take the usual precautions when using a microscope. Protect your eyes and skin from laboratory solutions. Be careful to avoid injury when using the glass probes. •

- ☐ 1 Obtain a small piece (1-2 cm) of glycerinated muscle tissue in a watch glass or petri dish.
- ☐ 2 Focus on the specimen after placing it on the stage of a dissecting microscope. (You may use a hand lens instead.)
- ☐ 3 Using fine glass probes or micropipette tips, try to separate some individual fibers from the rest of the bundle. Avoid using your hands or metal instruments to manipulate the specimen. Note: metal will "discharge" the tissue. Be sure to use glass!

> **Hint** ▶ Although it may seem difficult to separate fibers at first, care and patience will get the job done quickly and efficiently. Individual fibers are best, but a few in an unseparated bundle will do. •

- ☐ 4 Remove one of the fibers and place it on a microscope slide with a coverslip. Observe the specimen on a compound light microscope.

☐ 5 Observe the muscle fiber with low power, then high power. Make a sketch of your observation. Are the striations visible? Which parts of the muscle fiber form these stripes?

☐ 6 Put three or four separated fibers on a different microscope slide, straight and parallel to one another. Do not use a coverslip. With a metric ruler, measure the length of each fiber (in mm).

☐ 7 With the dropper built into the bottle, use the ATP + salt solution to bathe the fibers. Observe the fibers for 30 seconds and note any contraction.

☐ 8 Measure each fiber again and record the new length. Calculate the percentage of change in each fiber using the following formula:

$$\frac{(\text{Beginning Length}) - (\text{Ending Length})}{(\text{Beginning Length})} \times 100 = \%$$

What is the *average* percentage of change in the fibers? How do you account for the change? If there was no change, why not?

☐ 9 Repeat steps 6 through 8, using the ATP solution and again using the salt solution. Based on your observations, what chemicals are essential for contraction?

## B. CONTRACTION IN A MUSCLE ORGAN

Before beginning this activity, review these basic principles of muscle physiology:

The normal stimulus for a muscle fiber contraction is a *neurotransmitter* chemical from a nerve cell. The nerve-muscle fiber connection is called a **neuromuscular** junction. Electrical or mechanical stimulation can also cause a contraction. Regardless of whether the stimulus is chemical or electrical, it must be strong enough to pass the threshold of stimulation—the point at which it is just strong enough to initiate a contraction.

Fibers that compose a muscle organ are divided into functional teams, or motor units. Each fiber in a motor unit is innervated by the same nerve cell and so contracts at the same time. The relative strength of contraction in a whole organ depends in part on how many motor units are stimulated at one time. **Recruitment** of more units increases the organ's strength of contraction.

We will demonstrate muscle organ contraction by stimulating many motor units in a prepared muscle organ. To record our observations, we will use a **myograph**. Myograms, or graphic representations of muscle contractions, can be produced for isotonic contractions and for isometric contractions. Isotonic contractions change muscle length (without changing tension), and isometric contractions change tension (without changing length). On myograms, time is the horizontal axis, and strength (or length) of contraction is the vertical axis.

The myogram of a single **twitch** contraction shows these features (Figure 20-1, *A*):

☐ **Latent period**—Phase after stimulation has occurred but during which a contraction is not yet apparent

☐ **Contraction phase**—Period during which the myogram line rises, indicating that contraction is in progress
☐ **Relaxation phase**—Period when the line falls, indicating a return to the resting state

If restimulation of the muscle occurs soon enough after a twitch contraction, you may observe the **treppe** phenomenon, also called the *staircase effect* (see **Figure 20-1, *B***). The second in a series of contractions has a larger amplitude (size) than the first, the third an even greater amplitude, and so on. Eventually, a maximum is reached, and the waves reach a plateau. When the muscle can no longer sustain contractions, it is in a state of **muscle fatigue.**

When a muscle is continually restimulated before the relaxation phase is over, the myogram will not show the wave line returning to the baseline. Instead, the twitch waves will seem to fuse together. Wave summation, or fusion, is often termed **tetanus.** *Incomplete tetanus* occurs when a slight relaxation can be seen between waves. *Complete tetanus* occurs when the waves fuse into a line.

In this activity, you (or your instructor) will use the gastrocnemius muscle of a frog to demonstrate some of the major principles of muscle contraction.

Prepare a frog gastrocnemius muscle in the manner described in steps 1 through 7 and **Figure 20-2.**

> **⚠ safety first**
> Avoid cuts and punctures when using the dissection tools. Check for damaged or improper wiring on the electrical equipment. Do not touch the stimulator electrode or other bare wires directly. •

☐ 1 Holding the frog in your hand, locate the joint between the base of the skull and the first vertebra by bending the head down until a notch along the posterior edge of the skull can be felt or is visible.

☐ 2 Quickly push a pithing needle into the groove and then into the cranial cavity. While rotating the needle, move it from side to side to destroy the brain tissue inside.

☐ 3 Pull out the needle partially so that you can insert the tip into the vertebral canal. Push the needle as far as you can, again rotating it to destroy the nerve tissue.

☐ 4 Place the frog on a dissection pan or board. Lift the skin around the hip joint with a forceps and puncture it with the tip of a scissors. Try not to damage the muscle tissue of the leg.

☐ 5 Using the scissors, cut the skin in a circle all the way around the base of the thigh. Grasp the cut edge of the thigh's skin with forceps and pull it toward the foot. The skin will pull away from the leg as if it were a stocking. Keep pulling until the ankle joint appears.

> **Hint** Keep the muscle tissue moist with Ringer's solution so that it will remain functional. •

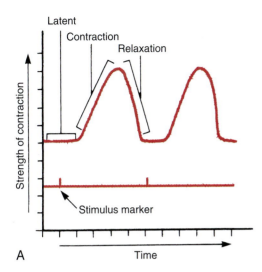

Latent

Contraction

Relaxation

Strength of contraction

Stimulus marker

A

Time

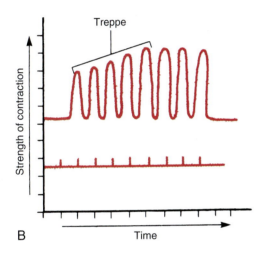

Treppe

Strength of contraction

B

Time

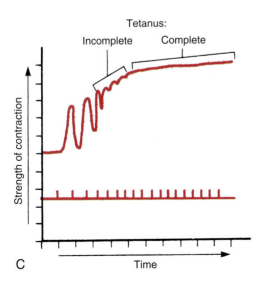

Tetanus:

Incomplete    Complete

Strength of contraction

C

Time

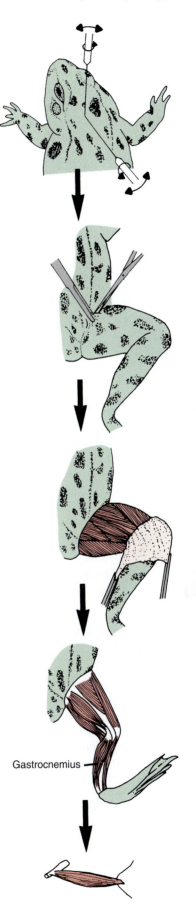

Gastrocnemius

**FIGURE 20-1**  A, Twitch contractions (two) on a myogram chart.
**B,** Treppe phenomenon. **C,** Tetanus.

**FIGURE 20-2**  Isolating frog gastrocnemius muscle.

☐ 6 Locate the *gastrocnemius muscle* (the fleshy part of the calf) and the *calcaneal* (Achilles) tendon, which attaches it to the ankle. Slip a blunt glass probe under the muscle, and slide it back and forth to detach the muscle from underlying connective tissue. While lifting the muscle with the probe, tie a 10- to 15-cm piece of string to the calcaneal tendon. Finally, cut the tendon distal to the knot.

☐ 7 Cut the thigh (bone and muscle) just proximal to the gastrocnemius muscle's attachment to the femur. Break the femur away from the knee joint and remove all muscles but the gastrocnemius. You now have a muscle ready to use in an experiment.

## HUMANE USE OF ANIMALS

Animals used in research and education should be treated respectfully and humanely. Double pithing is an accepted method for destroying the central nervous system so that the animal will feel no pain. This method also ensures that certain tissues will remain functional, including the neural tissue necessary for spinal reflexes. The frogs that you use have been raised for research and educational purposes.

☐ 8 Set up a kymograph or physiograph according to the instructions accompanying the instrument **(Figure 20-3)**. You may be measuring the intensity of an isotonic contraction by using a movable lever attached to the muscle (as in a kymograph). Or you may be measuring an isometric contraction by means of the force with which it pulls on a transducer (as in a physiograph).

> **Hint** ▶ Different systems operate differently, but here is an easy way to calibrate just about any physiograph setup:
>
> ☐ suspend a 100-g weight from the transducer
> ☐ adjust the signal amplifier SENSITIVITY or GAIN to produce a 5-cm deflection of the pen
>
> *Check the manufacturer's handbook for further details regarding your particular system.*

☐ 9 Set the stimulator *mode* to CONTINUOUS (REPEAT), *duration* (width) to 1 msec, *frequency* to 1 Hz (1 stimulus per second), and *voltage* to 0.1 V. Stimulate the muscle with the stimulator probe. Continue increasing the voltage in 0.1-V steps until a response is observed. This voltage is the *threshold of stimulation*. Record it here for later reference: _____ V.

☐ 10 Change the stimulator *mode* to SINGLE, then continue to increase voltage over a series of twitches a few seconds apart. You have to trigger each stimulus manually when in the SINGLE mode. Does the amplitude of the myogram curve remain the same? Why or why not? Record the voltage at which a moderately sized twitch is observed: _____ V.

☐ 11 Set the voltage to the level recorded in step 10, the *mode* to CONTINUOUS (REPEAT), and the *frequency* to 1 Hz. Keeping the voltage constant, record a series of twitches 1 second apart. Does the amplitude increase, decrease, or stay the same? Explain. Allow the muscle to rest for several minutes.

☐ 12 Keeping the voltage constant, increase the frequency slowly (from 1 Hz) until incomplete tetanus, then complete tetanus, is observed. Record the threshold frequencies for each phenomenon in the Lab Report at the end of the exercise.

## C. HUMAN MUSCLE CONTRACTION

This exercise may be done in addition to, or in place of, Activity B. Rather than using a laboratory animal as a subject, this activity calls for using *yourself* (or your lab partner) as an experimental subject. In this case, a muscle organ is used in vivo instead of being isolated from the subject's body. Due to the multitude of computer-assisted instruments that can measure electrical activity, this activity has been designed to be used with your instructor's choice from these instruments.

   **Please note:** Depending on the device, the following instructions may vary. Please check with your instructor for any modifications.

> ❗ *safety first*
>
> Do not participate in this activity as a subject if you have health problems that could be affected by this procedure (e.g., a heart condition). Be aware of electrical hazards. During the activity, the subject should not touch any person or object other than the device to test grip strength. •

☐ 1 Set up the instrument according to the user's manual provided by the manufacturer or as directed by your instructor.

☐ 2 Locate the **motor point** of the flexor digitorum superficialis muscle of the right forearm **(Figure 20-4)**. The motor point is the spot most sensitive to external stimulation. Use this method to find it:

   ☐ Put electrode gel on the flat plate electrode and attach it to the dorsal surface of the right hand with a rubber strap (tape).

   ☐ With the power switched off, set the stimulator *frequency* to 1 Hz (stimulus/sec), *mode* to continuous or repeat, *duration* to 1 msec, and *voltage* to 60 V. Spread a tiny dab of gel on the stimulator probe, and then turn the power on.

   ☐ Have the subject move the probe around the ventral surface of the forearm close to the belly of the flexor digitorum superficialis muscle until the area of strongest stimulation is found. When the strongest contraction of the third finger is observed, the subject has the correct spot. The subject may want to mark this spot with washable ink.

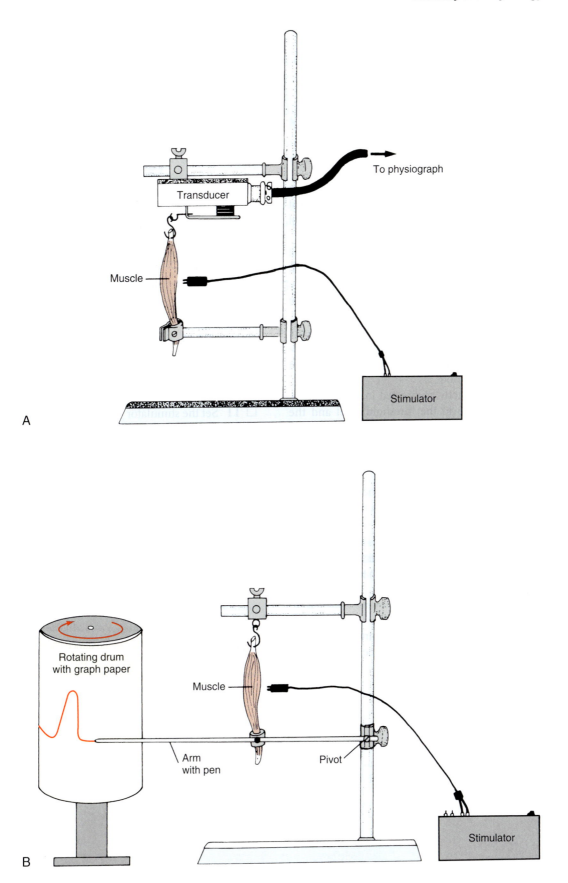

**FIGURE 20-3** **A,** Typical physiograph setup. **B,** Typical kymograph apparatus. Refer to your system's owner's manual, or your instructor, for detailed instructions in setting up the apparatus.

## Contraction in a Muscle Organ (provide the answer to each of the questions in the table below)

| PHENOMENON | VALUE | EXPLANATION |
|---|---|---|
| Threshold level of stimulation | V | What does this value represent? |
| Level of stimulation required for a standard twitch | V | What does this value represent? How does knowing this value help you do the rest of the experiment? |
| Series of twitch contractions | | Does the amplitude remain the same for each twitch in the series? Why or why not? |
| Threshold frequency or incomplete tetanus | Hz | What does this value represent? A common finding in the frog gastrocnemius is a threshold of about 70 Hz. Is your result higher or lower? |
| Threshold frequency or complete tetanus | Hz | What does this value represent? A common result is around 200 Hz. How does your result compare? |
| Duration of latent period (of one twitch) | msec | What does this value represent? Can you explain it in terms of the physiology of contraction? |
| Duration of contraction phase (of one twitch) | msec | What does this value represent? |
| Duration of relaxation phase (of one twitch) | msec | What does this value represent? |
| Total duration of one twitch contraction | msec | Looking at all your data, does this value seem to be constant or variable? |

NOTE: Do not forget to attach copies of your myogram charts (with your measurements and notations) to this report.

# Nerve Tissue

This exercise is the first of 11 that challenge you to investigate the human **nervous system.** Before going any further, it is best to lay a foundation by discussing the overall organization of this system.

The nervous system is composed of two major divisions: the **central nervous system (CNS)** and the **peripheral nervous system (PNS).** The CNS includes the brain and spinal cord; the PNS includes all the **nerves** that conduct impulses to and from the CNS. Often the nervous system is instead subdivided, based on function, into the **afferent division** and the **efferent division.** The afferent division includes nerves and tracts leading *toward* the CNS—the *sensory nerves* and *sensory tracts.* The efferent division includes nerves and tracts leading *away from* the CNS—the *motor nerves* and *motor tracts.*

The nervous system can be instead subdivided into the **somatic nervous system (SNS)** and the **autonomic nervous system (ANS).** The SNS is concerned mainly with pathways that monitor somatic sensors (general senses) and regulate somatic effectors (skeletal muscles). The ANS, on the other hand, mainly monitors and regulates subconscious visceral activities that involve cardiac muscle, smooth muscle, glands, and adipose tissue.

This exercise is an investigation of the cells of the nervous system. Lab Exercise 22—Nerve Reflexes will continue the study of the organization of the nervous system with an exploration of concepts of the *reflex arc.*

## BEFORE YOU BEGIN

- ☐ Read the appropriate chapter in your textbook and Lab Exercise 9—Muscle and Nerve Tissue in this Lab Manual.
- ☐ Set your learning goals. When you finish this exercise, you should be able to:
  - ☐ describe the structural components of a typical neuron and identify them on a model or chart
  - ☐ name the three structural categories of neurons and identify their principal locations
  - ☐ identify specimens of *multipolar* and *bipolar* neurons
  - ☐ identify a figure illustrating *unipolar* neurons
  - ☐ name the major categories of neuroglia and identify their principal locations
  - ☐ identify specimens of figures illustrating neuroglia
  - ☐ describe the structure of a typical peripheral nerve
- ☐ Prepare your materials:
  - ☐ models or charts:
    - ☐ multipolar neuron
    - ☐ unipolar neuron
    - ☐ nerve c.s.
  - ☐ compound light microscope
  - ☐ colored pencils or pens

- ☐ prepared microslides:
  - ☐ spinal cord smear (multipolar neurons)
  - ☐ retina c.s. (bipolar neurons)
  - ☐ peripheral nerve (myelinated) c.s.
  - ☐ spinal cord c.s.
- ☐ Read the directions and safety tips for this exercise *carefully* before starting any procedure.

> ⓘ *safety first*
> Observe the usual safety precautions when using the microscope. •

## Study TIPS

Before beginning this lab, review Lab Exercise 9—Muscle and Nerve Tissue. It will also be helpful if you have your textbook, the *A&P Survival Guide,* or images of neurons and each type of glial cell.

## A. THE NEURON

As you already know from your study of nerve tissue histology (Lab Exercise 9—Muscle and Nerve Tissue), the **neuron (Figure 21-1)** is the cell type that conducts impulses, or **action potentials.** The **neuroglia,** on the other hand, support the neurons in any number of ways. Neurons can be *unipolar,* having a single projection from the cell body; *bipolar,* having two projections from the cell body; or *multipolar,* having many projections (see Activity B). In this activity, you will find as many of the listed neuron structures as possible on a multipolar neuron model (or chart) and in a slide of multipolar neurons (spinal cord smear).

- ☐ 1 Locate the cell body, or **soma.** It is a large region made up of the nucleus and surrounding cytoplasm. Packets of rough endoplasmic reticulum (ER) in the soma produce neurotransmitters. These packets are still sometimes called **Nissl bodies** or *Nissl substance* after their discoverer, Franz Nissl.
- ☐ 2 The soma forms a cone-shaped projection, or **axon hillock,** as it projects to become the **axon.** The axon is one of two types of neuron projections (fibers). The axon usually conducts action potentials away from the cell body.
- ☐ 3 The axon may be wrapped with a series of neuroglial cells called **Schwann cells.** Schwann cells wrap like tape around the axons of some peripheral nerves, each spiraling around a fiber to form a multilayered coating. The inner layers of the Schwann cell are filled with the fatty white substance called **myelin.** Because the Schwann cells are found in series, they form a segmented *sheath of*

*Schwann*, or **myelin sheath.** The gaps between the Schwann cells are termed **nodes of Ranvier** *(myelin sheath gaps)*. The outer wrapping of each Schwann cell is normal cytoplasm, with organelles and a nucleus. In the case described, the axon is called a **white fiber,** or myelinated fiber. A group of white fibers together is called **white matter.** Schwann cells occur only in the PNS. Within the CNS, myelinated axons are wrapped with extensions of **oligodendrocytes,** another type of neuroglial cell.

☐ 4 Schwann cells (neurilemmocytes) or oligodendrocytes are also associated with **unmyelinated axons,** which together with cell bodies and dendrites form **gray matter.** However, in this case, the neuroglia do not form multiple wrappings and are not partially filled with myelin.

☐ 5 **Collateral axons,** or axon branches, can be observed in some cells. Also, the distal ends of axons are often branched. These smaller distal branches are termed **telodendria.**

☐ 6 Multipolar neurons have many projections from the soma called **dendrites.** Dendrites are branched extensions that are sensitive to stimuli from other cells. Other neurons form an association, or **synapse,** at a bump on the dendrite or cell body (soma). Stimulation of a dendrite or the cell body results in a local change in potential.

# B. STRUCTURAL CLASSIFICATION OF NEURONS

There are three categories of neurons based on their shapes **(Figure 21-2)**:

☐ 1 **Multipolar neurons**—Multipolar neurons have multiple projections from the cell body. Multipolar neurons comprise most of the neurons in the CNS. Almost all motor neurons are multipolar. Examine a multipolar neuron in a prepared spinal cord smear. Note the many projections.

☐ 2 **Bipolar neurons**—Bipolar neurons have exactly two projections from the cell body. Normally, one process conducts impulses toward the cell body. Such a process is called the *dendrite.* The other process conducts impulses away from the cell body and is called the *axon.* Bipolar neurons are most commonly found in the sensory tracts of the eye (for vision) and nasal epithelium (for smell). Examine a prepared slide of the retina (the sensitive lining of the eyeball). Can you distinguish any bipolar cells? A color micrograph of a retina cross-section is shown in **Figure 21-2.**

☐ 3 **Unipolar neurons**—Unipolar neurons, also called **pseudounipolar neurons,** have one extension from the cell body; however, the single process splits near the body to become two long branches. One branch has dendrite-like endings that receive stimulation for an impulse. The remainder of that branch and all of the other branches then conduct the impulse along. The two branches together function as a single axon. Most sensory neurons are unipolar. Examine a figure or model of a unipolar neuron.

## Coloring Exercises: Multipolar Neuron

Use colored pens or pencils to shade in both the figure and the labels. Each red numeral in the figure corresponds to a matching red numeral following the appropriate label.

SOMA 1
NUCLEUS 2
NUCLEOLUS 3
MITOCHONDRION 4
GOLGI APPARATUS 5
NISSL BODIES 6
AXON HILLOCK 7
DENDRITE 8
AXON 9
SCHWANN CELLS 10

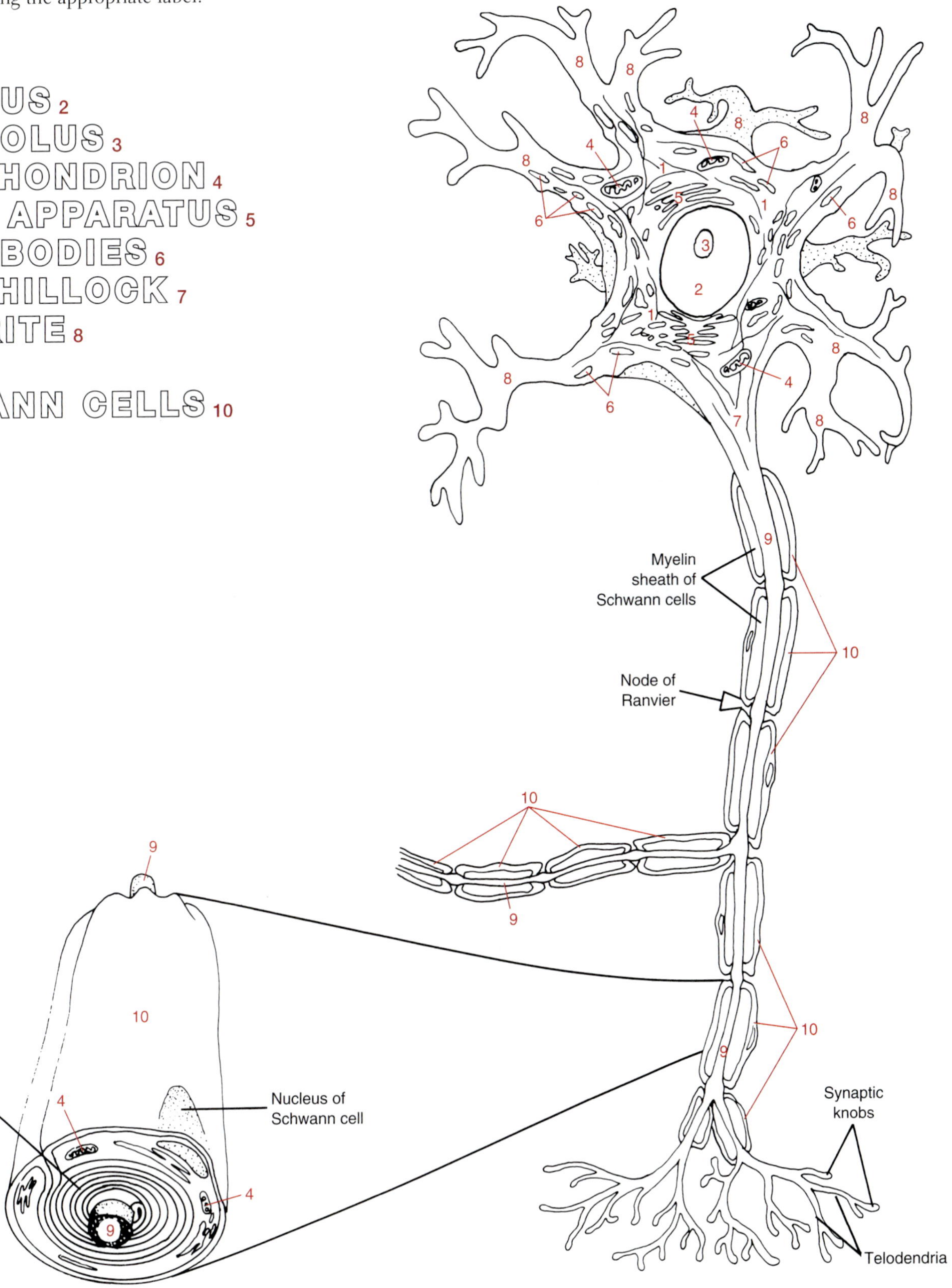

Myelin
sheath of
Schwann cells

Node of
Ranvier

Myelin
sheath

Nucleus of
Schwann cell

Synaptic
knobs

Telodendria

**FIGURE 21-1** Structure of a typical neuron.

Name: _____ Date: _____ Section: _____

# Nerve Tissue

Fill in this table summarizing your examination of a multipolar neuron model. Check off each structure as it is identified. For functions, refer to a reference book or your textbook.

| IDENTIFICATION | STRUCTURE | FUNCTION(S) |
|---|---|---|
| ☐ | Soma | |
| ☐ | Nissl bodies | |
| ☐ | Axon hillock | |
| ☐ | Axon | |
| ☐ | Schwann cells | |
| ☐ | Nodes of Ranvier | |
| ☐ | Collateral axon | |
| ☐ | Telodendria | |
| ☐ | Dendrite | |

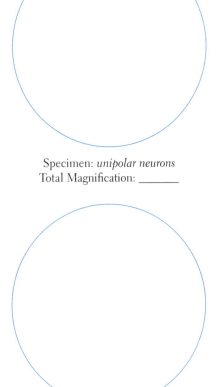

Specimen: *multipolar neurons*
Total Magnification: _____

Specimen: *unipolar neurons*
Total Magnification: _____

Specimen: *bipolar neurons*
Total Magnification: _____

Specimen: *Schwann cells*
Total Magnification: _____

> **Hint** The following structures may or may not be easily seen in the preserved specimen, depending on the specimen's condition. A freshly cut, even cross-section usually gives the best results. •

☐ 10 A cross-sectional view of the spinal cord shows a distinct H-shaped area of gray matter surrounded by areas of white matter. The two lateral sections of gray matter are joined by a transverse **gray commissure.** In the center of the gray commissure, you may see the **central canal.** This canal contains CSF in life but contains preservative in prepared specimens.

☐ 11 Each lateral section of the gray matter exhibits extensions termed the **anterior gray horn** and the **posterior gray horn.** The thoracic and lumbar regions also have lateral extensions called **lateral gray horns** and contain neuron cell bodies of interneurons and motor neurons as do the anterior and posterior gray horns.

☐ 12 Columns of white matter surround the central gray matter area. These include the **anterior white column, lateral white column,** and **posterior white column** on each side. Columns are sometimes called **funiculi.**

> ❗ **safety first**
> Observe the usual precautions when using the microscope and slides. •

## B. MICROSCOPIC EXPLORATION

☐ 1 Obtain a stained cross-section preparation of a mammalian spinal cord and observe it at a low total magnification.

☐ 2 Try to find as many features of the spinal cord (as described in Activity A) as you can. White matter and gray matter are often distinguished by differential staining (contrasting colors), as in **Figure 21-4.**

## C. SPINAL NERVES

Although not really part of the spinal cord proper (they are in the PNS rather than the CNS), we will consider spinal nerves in this activity. As each structure is presented, try to locate it on a model or chart of the human nervous system and in **Figure 23-2.**

☐ 1 There are 31 pairs of spinal nerves communicating with the spinal cord by way of the intervertebral foramina formed as the vertebrae stack on one another. The spinal nerves are named for the vertebral region with which they are associated:
  ☐ **Cervical spinal nerves**—C1 through C8 (one pair between the skull and vertebra C1 and one pair inferior to each of seven cervical vertebrae)
  ☐ **Thoracic spinal nerves**—T1 through T12
  ☐ **Lumbar spinal nerves**—L1 through L5
  ☐ **Sacral spinal nerves**—S1 through S5 (branches of these nerves communicate via the dorsal and pelvic foramina of the sacrum)
  ☐ **Coccygeal spinal nerves**—C or Co

☐ 2 Each spinal nerve communicates with the cord by means of two separate **spinal nerve roots,** each formed by the combination of six to eight tiny *rootlets.* The **ventral (anterior) spinal nerve root** includes efferent fibers (somatomotor and autonomic axons). The **dorsal (posterior) spinal nerve root** contains afferent fibers. Cell bodies of afferent neurons form an area of gray matter in the dorsal root called the **dorsal root ganglion.** The spinal nerve roots are found within the spinal cavity. The spinal nerve proper passes through the intervertebral foramina.

☐ 3 Outside the spinal cavity, each spinal nerve branches. The largest branches are the **ventral ramus** and the **dorsal ramus.** In most of the thoracic and lumbar nerves, the ventral ramus branches into autonomic and somatic pathways. In other segments, the fibers of the ventral rami branch to form **plexi.** Plexi are complex nerve networks in which fibers from several different spinal nerves are reorganized to form specific peripheral nerves. Identify each plexus on a model or chart:
  ☐ Cervical plexus
  ☐ Brachial plexus
  ☐ Lumbar plexus
  ☐ Sacral plexus

# Coloring Exercises: Spinal Cord

Use colored pens or pencils to shade in both the figure and the labels. Each red numeral in the figure corresponds to a matching red numeral following the appropriate label.

**Meninges**
DURA MATER 1
ARACHNOID MATER 2
SUBARACHNOID SPACE 3
PIA MATER 4

**Spinal Cord**
CENTRAL CANAL 5
GRAY COMMISSURE 6
ANTERIOR GRAY HORN 7
POSTERIOR GRAY HORN 8
ANTERIOR WHITE COLUMN 9

POSTERIOR WHITE COLUMN 10
LATERAL WHITE COLUMN 11

**Spinal Nerve**
VENTRAL NERVE ROOT 12
DORSAL NERVE ROOT 13
DORSAL ROOT GANGLION 14
SPINAL NERVE PROPER 15
VENTRAL RAMUS 16
DORSAL RAMUS 17

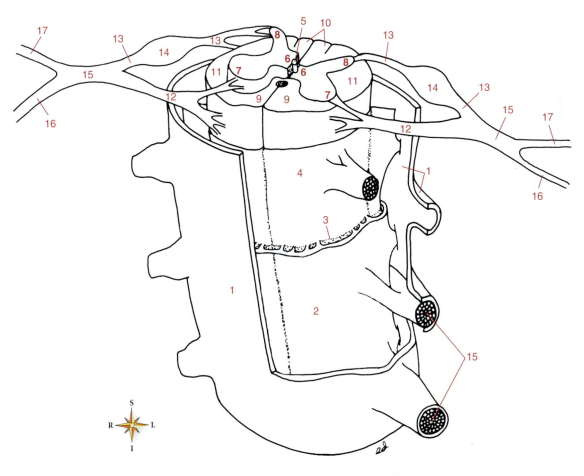

**FIGURE 23-1** Structure of the spinal cord.

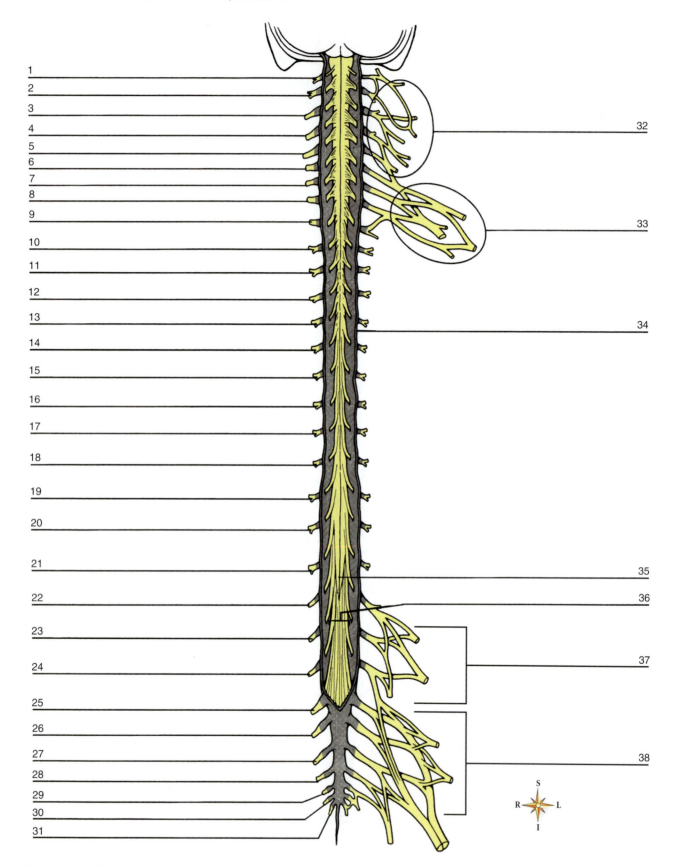

**FIGURE 23-2** In the blanks numbered 1 through 31, write the correct spinal nerve abbreviation. In blanks 32 through 38, write the name of the nerve feature indicated. Also, write the abbreviations and nerve features on the corresponding blanks on the Lab Report at the end of this exercise.

Name: _____  Date: _____  Section: _____

# LAB REPORT 23
# The Spinal Cord and Spinal Nerves

**Figure 23-2** (from p. 254)

1. _____
2. _____
3. _____
4. _____
5. _____
6. _____
7. _____
8. _____
9. _____
10. _____
11. _____
12. _____
13. _____
14. _____
15. _____
16. _____
17. _____
18. _____
19. _____
20. _____
21. _____
22. _____
23. _____
24. _____
25. _____
26. _____
27. _____
28. _____
29. _____
30. _____
31. _____
32. _____
33. _____
34. _____
35. _____
36. _____
37. _____
38. _____

Summarize your exploration of the spinal cord and spinal nerves by using this checklist.

| STRUCTURE | NOTES |
|---|---|
| ☐ dura mater | |
| ☐ arachnoid mater | |
| ☐ subarachnoid space | |
| ☐ pia mater | |
| ☐ anterior median fissure | |
| ☐ posterior median sulcus | |
| ☐ cervical enlargement | |
| ☐ lumbar enlargement | |
| ☐ medullary cone | |
| ☐ cauda equina | |
| ☐ ventral spinal nerve root | |
| ☐ dorsal spinal nerve root | |
| ☐ dorsal root ganglion | |
| ☐ gray commissure | |
| ☐ central canal | |
| ☐ gray horns:<br>　☐ anterior<br>　☐ lateral<br>　☐ posterior | |
| ☐ white columns:<br>　☐ anterior<br>　☐ lateral<br>　☐ posterior | |
| ☐ spinal nerves:<br>　☐ cervical<br>　☐ thoracic<br>　☐ lumbar<br>　☐ sacral<br>　☐ coccygeal | |
| ☐ ventral ramus of spinal nerve | |
| ☐ dorsal ramus of spinal nerve | |

**Put in Order I**

1. _____
2. _____
3. _____
4. _____
5. _____
6. _____

**Put in Order II**

1. _____
2. _____
3. _____
4. _____
5. _____

**Multiple Choice**

1. _____
2. _____
3. _____
4. _____
5. _____

**Put in Order I** (rearrange these structures in the order in which a sensory signal might pass)

cervical plexus
dorsal root ganglion
posterior gray horn
receptor
spinal nerve C2
ventral ramus

**Put in Order II** (rearrange these structures in the order in which a motor signal might pass)

anterior gray horn
effector
spinal nerve T6
ventral ramus
ventral nerve root

**Multiple Choice** (only one response is correct in each item that follows)

1. The medial deep groove on the ventral surface of the spinal cord is called the
   a. posterior median sulcus
   b. medial enlargement
   c. anterior lateral sulcus
   d. anterior median fissure
   e. ventral lateral column
2. The correct order for the meninges of the brain (from superficial to deep) is
   a. pia mater, dura mater, arachnoid
   b. dura mater, arachnoid, pia mater
   c. dura mater, pia mater, arachnoid
   d. dura mater, subarachnoid, pia mater
   e. the brain does not have meninges
3. The cauda equina is
   a. not present in humans
   b. a cone-shaped end of the spinal cord proper
   c. a bundle of separate spinal nerve roots
   e. a bundle of cranial nerves
   f. an enlargement of the spinal cord proper
4. Spinal nerve branches from several spinal segments (vertical levels) may exchange fibers. Such a network of nerve fibers is called a
   a. braid
   b. ramus
   c. plexus
   d. ganglion
   e. spinal nerve root
5. Large groups of neuron cell bodies are likely to be found in
   a. the lateral gray horn
   b. the gray commissure
   c. a dorsal root ganglion
   d. the central canal
   e. a, b, and c are correct

# 24
# The Brain and Cranial Nerves

Use col
numera

**Lobe**

FRO
PAR
OCC
TEM
INS

**Func**

PRI
PRI
MO
PRI
M

Having explored the spinal cord in the previous exercise, it is time to move on to an exploration of the other major organ of the central nervous system: the **brain.** The brain occupies most of the cranial cavity within the skull. This rather large, complex organ has been partitioned, for convenience of study, into numerous divisions and subdivisions. Nerves that enter and exit the brain directly are called **cranial nerves.** In this exercise, you are challenged to explore the organization of the brain and the structure and function of the cranial nerves.

## BEFORE YOU BEGIN

- ☐ Read the appropriate chapter in your textbook.
- ☐ Set your learning goals. When you finish this exercise, you should be able to:
  - ☐ list and describe the principal structures of the brain
  - ☐ identify structures of the human brain in a chart, model, or preserved specimen
- ☐ Prepare your materials:
  - ☐ chart, model, or preserved specimen of a human brain
  - ☐ colored pencils or pens
  - ☐ computer setup with a program to simulate dissection of a human body (optional)
  - ☐ protective gear: gloves, safety eyewear, and apron
- ☐ Read the directions for this exercise *carefully* before starting any procedure.

### *Study* TIPS

Obtain an unlabeled illustration/photo of the external brain. You may want to use your cell phone to take a picture. Print this picture. As you complete this lab, use the unlabeled illustration/photo and label the external aspects of the brain.

> **Hint** Your textbook has descriptions and figures that will help you identify structures of the brain. You may wish to use your textbook and the *Brief Atlas* as supplemental resources for this and the following activities. •

## A. EXTERNAL ASPECT OF THE BRAIN

Using a model or preserved specimen of the human brain, locate the structures of the brain listed in the following steps.

- ☐ 1 Determine whether any of the **meninges,** or membranous coverings of the brain, are intact **(Figure 24-1)**. If they are present, locate the following:
  - ☐ **Dura mater**
  - ☐ **Sagittal sinus**
  - ☐ **Subdural space**
  - ☐ **Arachnoid mater**
  - ☐ **Subarachnoid space**
  - ☐ **Pia mater**
- ☐ 2 Locate the largest part of the brain, the **cerebrum.** The cerebrum is divided into nearly symmetrical left and right **cerebral hemispheres** by a deep **longitudinal fissure.** Joining the left and right cerebral hemisphere is a band of white matter called the **corpus callosum,** which may be seen at the bottom of the longitudinal fissure. The surface of the cerebrum, or **cerebral cortex,** is characterized by large convoluted folds of gray matter called **gyri** (singular, gyrus). The grooves between the gyri are called **sulci** (singular, sulcus) if they are shallow and **fissures** if they are deep. Some fissures serve as landmarks to divide the cerebrum into regions called **lobes** that correspond to the overlying skull bones:
  - ☐ **Frontal lobe** (left and right)
  - ☐ **Parietal lobe**
  - ☐ **Occipital lobe**
  - ☐ **Temporal lobe**
  - ☐ **Insula** (also called the *island of Reil,* a region hidden within a fissure that separates the temporal lobe from the parietal lobe)

In addition to using lobes as a means of identifying different regions of the cerebral cortex, neurobiologists often prefer to identify **functional regions** of the cortex. **Figure 24-2** is a coloring exercise that illustrates some of the major

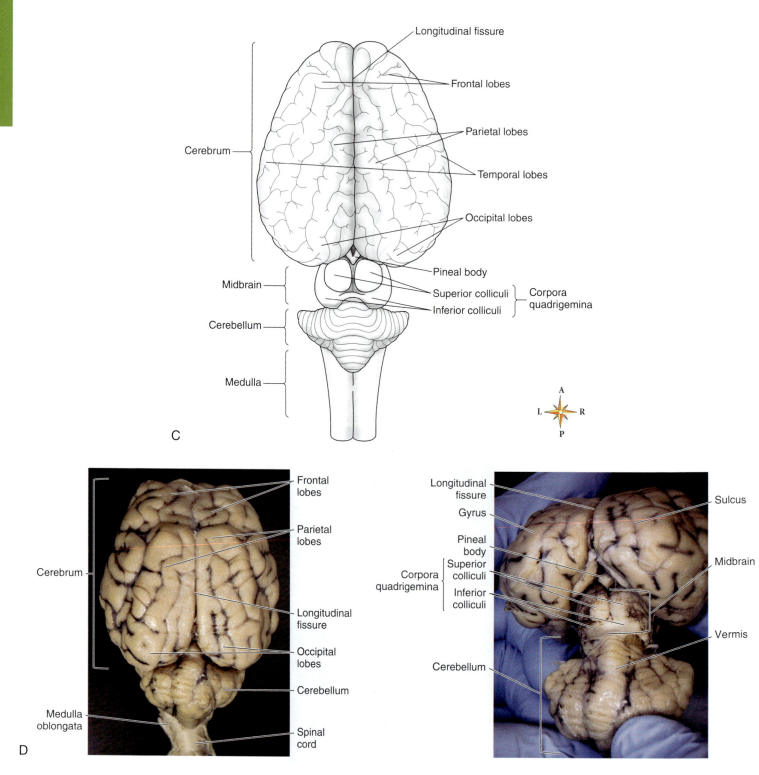

**FIGURE 25-1, cont'd**  C, Dorsal aspect. D, Dorsal view photographs.

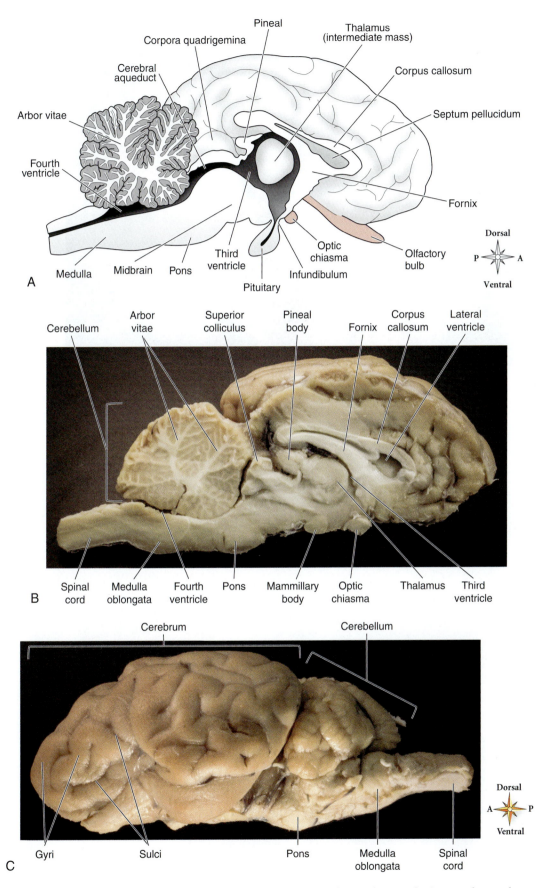

FIGURE 25-2 Sheep brain, left side. A, Midsagittal section. B, Midsagittal view photograph. C, Lateral view photograph.

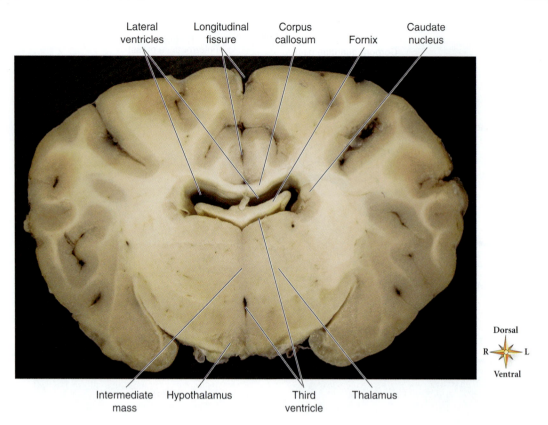

Lateral ventricles   Longitudinal fissure   Corpus callosum   Fornix   Caudate nucleus

Dorsal
R ← → L
Ventral

Intermediate mass   Hypothalamus   Third ventricle   Thalamus

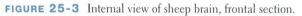

**FIGURE 25-3** Internal view of sheep brain, frontal section.

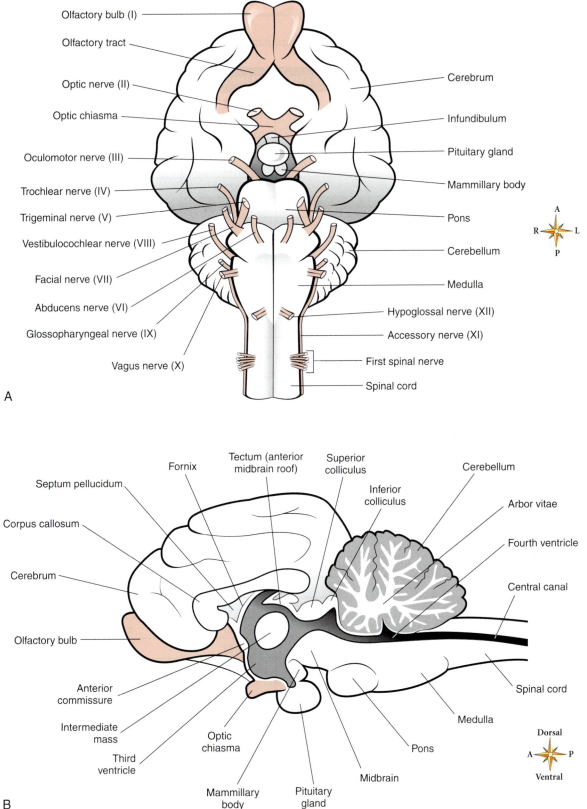

**A**

- Olfactory bulb (I)
- Olfactory tract
- Optic nerve (II)
- Optic chiasma
- Oculomotor nerve (III)
- Trochlear nerve (IV)
- Trigeminal nerve (V)
- Vestibulocochlear nerve (VIII)
- Facial nerve (VII)
- Abducens nerve (VI)
- Glossopharyngeal nerve (IX)
- Vagus nerve (X)
- Cerebrum
- Infundibulum
- Pituitary gland
- Mammillary body
- Pons
- Cerebellum
- Medulla
- Hypoglossal nerve (XII)
- Accessory nerve (XI)
- First spinal nerve
- Spinal cord

A
R ← → L
P

**B**

- Septum pellucidum
- Corpus callosum
- Cerebrum
- Olfactory bulb
- Anterior commissure
- Intermediate mass
- Third ventricle
- Mammillary body
- Optic chiasma
- Pituitary gland
- Fornix
- Tectum (anterior midbrain roof)
- Superior colliculus
- Inferior colliculus
- Cerebellum
- Arbor vitae
- Fourth ventricle
- Central canal
- Spinal cord
- Medulla
- Pons
- Midbrain

Dorsal
A ← → P
Ventral

**FIGURE 25-4** Cat brain. **A,** External view of ventral aspect. **B,** Midsagittal section.

# Somatic Senses

Sensations occur in the central nervous system (CNS) after sensory stimuli in the internal or external environment generate nerve impulses, which in turn convey information regarding the environmental change to the CNS. Sensory receptors can be classified in a number of ways. They can be classified by **mode** (e.g., *photoreceptors* are sensitive to light, *chemoreceptors* are sensitive to certain chemicals, and *mechanoreceptors* are sensitive to mechanical changes). Sensory receptors can also be classified according to their **location** (e.g., *exteroceptors* are located in the skin, *proprioceptors* are located in the muscles, and *visceroceptors* are located in deeper tissues [the viscera]). Senses mediated by the sensory receptors of the body can be categorized into two major groups: the **somatic senses** and the **special senses.** The special senses involve elaborate sensory organs such as the eye or ear. Somatic senses involve less elaborate sensory mechanisms, often just single receptors embedded in skin or muscle tissue. This exercise briefly explores some of the somatic senses. Later exercises discuss the special senses.

## BEFORE YOU BEGIN

☐ Read the appropriate chapter in your textbook.
☐ Set your learning goals. When you finish this exercise, you should be able to:
  ☐ explain the distinction between *somatic senses* and *special senses*
  ☐ demonstrate that exteroceptors are not evenly distributed in the skin
  ☐ demonstrate the use of the proprioceptive sense in controlling body movement
☐ Prepare your materials:
  ☐ compass
  ☐ metric ruler (millimeter)
  ☐ pen (washable ink) or rubber stamp
  ☐ horse hair
  ☐ blunt metal probe
  ☐ hot water tap
  ☐ ice water
  ☐ chalkboard and chalk, or whiteboard and dry erase marker
☐ Read the directions and safety tips for this exercise *carefully* before starting any procedure.

## Study TIPS

To help you understand the classification of the sensory receptors, develop a concept map. For reference, use your textbook or the *A&P Survival Guide*. Develop the map to include the mode and location of these sensory receptors.

## A. CUTANEOUS SENSES

Cutaneous senses are somatic senses that are mediated by exteroceptors—sensory receptors in the skin. Cutaneous senses include sensitivity to touch, pressure, heat, cold, vibration, and any other stimulus felt through the skin. In this activity, you will explore the distribution pattern of some of the cutaneous senses.

☐ 1 Perform the two-point discrimination test **(Figure 26-1)** on your lab partner by following these directions:
  ☐ Adjust so that two points are touching.
  ☐ With your partner's eyes closed and hand placed palm upward on the lab table, gently place the points on the surface of your partner's index finger.
  ☐ Move the points apart by about 1 mm and touch the finger again.
  ☐ Repeat this sequence, each time increasing the distance slightly, until your partner tells you that two distinct points are felt. Record the distance between points in the lab report.
  ☐ Repeat the test on the back of the arm and compare the density of cutaneous touch receptors in these two external body regions.

> ⓘ **safety first**
> Do not press hard with the compass. Pressing hard ruins the results of the test and may cause injury to the subject.  •

☐ 2 Measure the density of touch receptors on the back of the hand using a different method:
  ☐ Draw a 2.5-cm (1-inch) square on the back of your lab partner's hand with a nontoxic, washable marker or rubber stamp. Divide the square into 16 blocks by drawing a grid within the box (use a ruler).
  ☐ With your partner's eyes closed, touch the tip of a horse hair to a point on the grid just hard enough to bend it slightly.
  ☐ Mark the position on the matching grid in the Lab Report at the end of this exercise. Record "0" if the subject does not feel the touch; record "+" if the subject does feel the touch. Repeat this for about 25 to 35 different points within the grid.

> **Hint** ▶ Touch your partner at random intervals—not with a noticeable rhythm—to avoid anticipation that might alter your results. Also, do not move from block to block in an orderly pattern. This might also cause the subject to anticipate your next move and indicate that he or she felt the hair even though he or she did not really feel it.  •

☐ 3 Adaptation can occur in many types of sensory receptors, including odorant receptors. In adaptation, prolonged exposure to a particular stimulus causes a receptor to fatigue and thus cease sending information to the central nervous system. Olfactory adaptation is common (e.g., when we put on cologne or use a scented soap and soon cannot smell it on ourselves). Demonstrate the concept of sensory adaptation by performing the following activity:

    ☐ Hold one of the vials used in step 2 near your nose, breathing normally, but still able to smell the substance.

    ☐ Have your partner begin timing (using a stopwatch or timer) as soon as you begin smelling the substance.

    ☐ Continue to hold the vial in place for several minutes, indicating to your partner the moment you can no longer smell the odor distinctly.

    ☐ Record your adaptation time in the Lab Report at the end of this exercise.

## B. SENSE OF TASTE

The sense of taste includes at least five modalities of chemical sensation: *salt, sweet, sour, bitter,* and *umami (glutamate)*. The receptor organs that mediate these sensations are the **taste buds,** found primarily along the sides of the papillae of the tongue. **Gustatory hairs** projecting from sensory cells possess receptors that can be triggered by the molecules called **tastants** detected by the sense of taste.

> **Hint** ▸ Color micrographs of the tongue papillae and taste buds can be found in **Figures 27-2** and **27-3**. •

☐ 1 Locate taste buds embedded within the papillae of a prepared microscopic specimen of the human tongue. Sketch your observations in the Lab Report at the end of this exercise.

You are already familiar with the idea of tastes such as salty, sweet, sour, and bitter. You may not be aware of the word *umami,* but you surely have tasted it. Umami is a Japanese word that describes the taste of glutamate, an amino acid found in many foods. Flavor enhancers such as MSG (e.g., Accent) add umami flavors to food. Meats and tomato products such as ketchup have a strong umami taste.

☐ 2 Determine the sensitivity of the various types of taste buds (salt, sweet, sour, bitter, and umami) by following this procedure:

    ☐ Blot the saliva from the surface of your tongue with a *clean* paper towel.

    ☐ Dip a sterile, disposable cotton swab in one of the five taste solutions provided in a clean cup and place it on various areas of the tongue. Record whether you can perceive the taste by checking the list provided in the Lab Report at the end of this exercise.

    ☐ Repeat this procedure for the remaining taste solutions:

        ☐ *salt:* salt (NaCl) solution

        ☐ *sweet:* sugar solution

        ☐ *sour:* lemon juice

        ☐ *bitter:* bitter almond solution

        ☐ *glutamate (umami):* MSG solution

> ⓘ **safety first**
>
> Use only *fresh,* uncontaminated solutions. Do not reuse the cotton swabs, even with the same person. Never dip a used swab into a solution. Dispose of used swabs properly. Check the contents of each solution, and avoid contact with substances that your physician has advised you to avoid or to which you think you might be overly sensitive. •

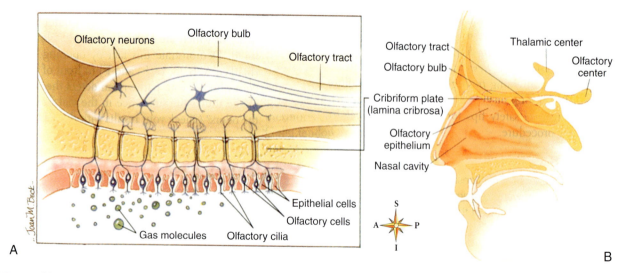

**FIGURE 27-1** Olfactory epithelium. **A,** Enlarged view of cross-section of olfactory epithelium and olfactory bulb that supplies sensory fibers to it. **B,** Location of olfactory epithelium in nasal cavity.

Stratified squamous epithelium

Vallate
papillae

Taste buds

Moat

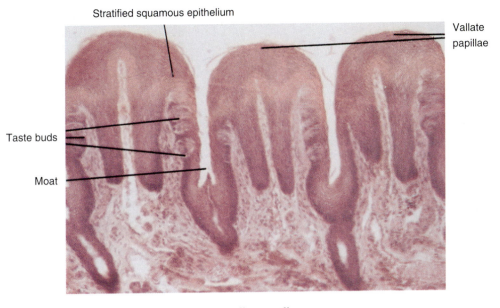

**FIGURE 27-2** Vallate papillae on tongue.

Papilla     Taste bud

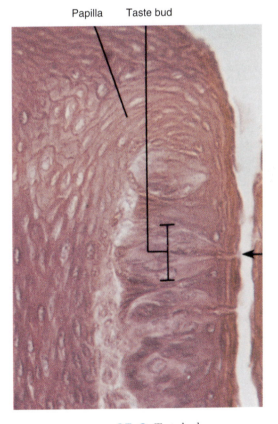

**FIGURE 27-3** Taste buds.

Name: _____   Date: _____   Section: _____

**LAB REPORT 27**
# Senses of Taste and Smell

## Olfactory Recognition

| VIAL | UNRECOGNIZED | RECOGNIZED | SUBSTANCE |
|------|--------------|------------|-----------|
| A | ☐ | ☐ | |
| B | ☐ | ☐ | |
| C | ☐ | ☐ | |
| D | ☐ | ☐ | |
| E | ☐ | ☐ | |
| F | ☐ | ☐ | |

## Olfactory Recognition

| TRIAL | SUBSTANCE | ADAPTATION TIME |
|-------|-----------|-----------------|
| 1 | | |
| 2 | | |

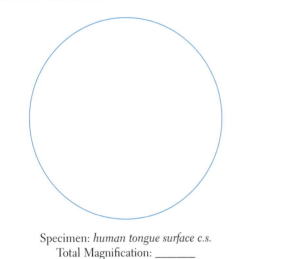

Specimen: *human tongue surface c.s.*
Total Magnification: _____

**Taste bud modality**
☐ salt
☐ sweet
☐ sour
☐ bitter
☐ umami

**Essay:** Write a few brief paragraphs outlining how the senses of smell and taste operate in ways that help the body maintain a homeostatic balance. Include at least two examples of how each might be involved in a homeostatic feedback loop. Use another sheet of paper if you need more space.

# The Ear

The **ears** are complex sensory organs located on each side of the skull, mostly buried within each temporal bone. The ear serves as a receptor organ for at least three special senses: *hearing, static equilibrium,* and *dynamic equilibrium.* This exercise challenges you to learn the basic divisions and structures of the human ear. Once you are familiar with the structure of the ear, you will be ready to explore the functional aspects of hearing and equilibrium in Lab Exercise 29—Hearing and Equilibrium.

## BEFORE YOU BEGIN

☐ Read the appropriate chapter in your textbook.
☐ Set your learning goals. When you finish this exercise, you should be able to:
   ☐ distinguish the three main divisions of the ear
   ☐ identify the principal structures of the human ear in models and figures
   ☐ locate features of the cochlea in a prepared microscopic specimen
☐ Prepare your materials:
   ☐ model or chart of the human ear
   ☐ compound light microscope
   ☐ prepared microslide: *cochlea c.s.*
   ☐ colored pencils or pens
☐ Read the directions and safety tips for this exercise *carefully* before starting any procedure.

### *Study* TIPS

Develop a chart that summarizes the different structures of the ear. It would be helpful if you also categorize them according to the division of the ear in which they are located (external, middle, or inner). You may also want to develop a concept map that illustrates the order of these structures in which sound waves pass through them. For reference, use your textbook or the *A&P Survival Guide*.

> **Hint** ▶ Use a chart or model of the human ear to locate and study the ear structures listed in the three activities of this exercise. •

> **!** *safety first*
> Observe the usual precautions when using the microscope and prepared slides. •

## A. EXTERNAL EAR

You may want to use your textbook as an additional aid in this activity. The **external (outer) ear** is the ear division composed of the elements described in the following:

☐ 1 **Auricle (pinna)**—This external ear flap protects the auditory opening and directs sound waves toward it. It also functions as a "radiator" in thermoregulation.
☐ 2 **External auditory meatus**—This tubelike passage carries airborne sound waves further into the ear apparatus; it is also called the *ear canal.*
☐ 3 **Tympanic membrane**—Also called the *eardrum,* it covers the end of the external auditory meatus to form a boundary with the middle ear. It vibrates when struck by airborne sound waves, carrying the sound energy into the middle ear.

## B. MIDDLE EAR

The **middle ear** begins where the external ear ends, with the tympanic membrane. The middle ear is an air-filled cavity lined with mucous membrane. Identify the important parts of the middle ear described in the following:

☐ 1 **Malleus**—It is one of the three *auditory ossicles* in each ear. Also called the *hammer,* it is a tiny, club-shaped bone attached to the eardrum. It vibrates when sound waves pass to it from the eardrum.
☐ 2 **Incus**—Also called the *anvil,* this tiny bone forms a synovial joint with the malleus. The incus vibrates when it receives energy from the malleus.
☐ 3 **Stapes**—Also called the *stirrup* because of its shape, this ossicle is joined to the incus, from which it receives vibrations. A flat portion of the stapes fits into the **oval window,** a passage into the inner ear. Because of the structural relationship of this chain of ossicles, sound waves are carried from the tympanic membrane to the oval window (that is, from the external ear to the inner ear). Along the way, the vibrations are amplified for better reception.
☐ 4 **Auditory (eustachian) tube**—This is a collapsible tube running between the middle ear and the pharynx. It allows internal air pressure to equalize with atmospheric air pressure so that high pressure on one side does not distort or muffle the eardrum.
☐ 5 The middle ear also has an opening to the mastoid air cells in the mastoid process of the temporal bone. What significance might this have if the middle ear becomes infected?

# C. INNER EAR

The **inner ear** is the third division of the ear apparatus. The receptors for hearing and equilibrium are located here. The inner ear is within a hollow area in the petrous portion of the temporal bone. A mazelike **bony labyrinth** contains a similarly shaped but smaller **membranous labyrinth.** The fluid inside the membranous labyrinth is called **endolymph,** and the fluid around the outside the membranous labyrinth is called **perilymph.** The bony labyrinth, and the membranous labyrinth inside it, is composed of the three main regions described in the following:

☐ 1 Cochlea—This is a long passage coiled like a snail. A cross-section reveals that the passage is divided into three chambers by a Y-shaped partition. The base of the Y is a projection of the bone called the **spiral lamina.** The two branches are pieces of membrane called the **vestibular membrane** and the **basilar membrane.** The space between the two membranes is the endolymph-filled **cochlear duct.** The space outside the vestibular membrane is the **scala vestibuli,** whereas the **scala tympani** is the space outside the basilar membrane. Sound waves move into the perilymph of the scala vestibuli as the stapes vibrates in the oval window. The vestibular membrane vibrates, causing the endolymph in the cochlear duct to vibrate. This, in turn, causes the flaplike **tectorial membrane** to vibrate and bend the hairs projecting from the **organ of Corti** (spiral organ) on the basilar membrane (which also vibrates). The bending of hairs induces receptor potentials in the sensory neurons. The energy dissipates as it moves through the scala tympani to the round window. A branch of cranial nerve VIII called the **cochlear nerve** carries information to auditory areas in the brain. Observe a cross-section of a mammalian cochlea in the microscope and sketch your observations. Label as many parts as you can identify.

**Hint** ▶ **Figure 28-3** shows a light micrograph of the mammalian cochlea and features a good view of the organ of Corti in cross-section. ●

☐ 2 **Vestibule**—This is the central area of the inner ear. The vestibule contains saclike portions of the membranous labyrinth called the **utricle** and the **saccule.** Each has a patch of sensory hair cells called a **macula (Figure 28-2, A).** The macula's hair cells are covered by a gelatinous coating embedded with hard, tiny crystals called **otoliths.** When the head is tilted, gravity pulls on the heavy otoliths and the hairs bend. This induces a receptor potential. Sensory information regarding the effects of gravity (static equilibrium) is transmitted to the brain through the **vestibular nerve,** another branch of cranial nerve VIII.

☐ 3 **Semicircular canals**—These are three round passages, each on a different plane. The semicircular canals have bubbles at their bases called **ampullae** (see **Figure 28-2, B).** Within each ampulla is a crest of tissue called the **crista ampullaris.** Each crista is a patch of sensory hair cells covered with a gelatinous mass (without otoliths) called the **cupula.** When the speed or direction of movement of the head changes, the inertia of the endolymph within the semicircular canals causes it to circulate. As the endolymph circulates, it pushes the cupula and generates receptor potentials in the crista's sensory cells. The vestibular nerve carries the signal to the brain, where it is interpreted as kinetic (dynamic) equilibrium **(Figure 28-1).**

## Coloring Exercises: Ear

Use colored pens or pencils to shade in both the figure and the labels. Each red numeral in the figure corresponds to a matching red numeral following the appropriate label.

### External Ear
AURICLE 1

EXTERNAL AUDITORY MEATUS 2

TYMPANIC MEMBRANE 3

### Middle Ear
MALLEUS 4

INCUS 5

STAPES 6

AUDITORY TUBE 7

### Inner Ear
COCHLEA 8

SPIRAL LAMINA 9

VESTIBULAR MEMBRANE 10

BASILAR MEMBRANE 11

COCHLEAR DUCT 12

SCALA VESTIBULI 13

SCALA TYMPANI 14

TECTORIAL MEMBRANE 15

ORGAN OF CORTI 16

COCHLEAR NERVE 17

VESTIBULE 18

VESTIBULAR NERVE 19

SEMICIRCULAR CANALS 20

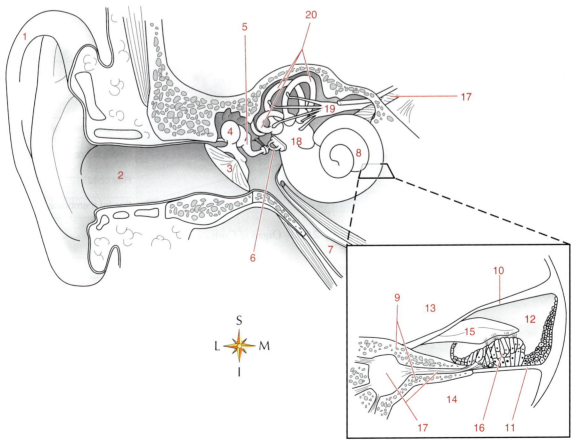

**FIGURE 28-1** External, middle, and inner structures of the ear.

**macula lutea** with a pit called the **fovea centralis.** The fovea is normally the center of the visual field and contains many cones. Rods become more dominant farther away from the fovea. Medial to the macula lutea is the white **optic disk.** Here, blood vessels, as well as the nerve fibers that exit as the **optic nerve,** pass out of the eyeball.

☐ Observe a cross-section of the retina in a prepared microscopic specimen. Can you identify the features of the retina shown in **Figure 30-2**? Are either of the other two coats of the eyeball visible in your specimen?

> **Hint** ▶ **Figure 30-2,** *C*, shows a light micrograph of the retina. •

☐ 5  Identify these cavities and chambers of the eye:
  ☐ **Anterior cavity**—Filled with **aqueous humor,** a watery filtrate produced by the ciliary body. Aqueous humor circulates from the **posterior chamber** behind the iris, through the pupil, and to the **anterior chamber,** where it is reabsorbed.
  ☐ **Posterior cavity**—Filled with transparent, jellylike **vitreous body.** This cavity is posterior to the lens.

## B. EYE DISSECTION

Apply your knowledge of mammalian eye anatomy by carefully dissecting a cow or sheep eye (**Figures 30-4** and **30-5**).

> **⚠ safety first**
> Observe the usual precautions while dissecting the preserved or fresh tissue of the eye specimen. Use safety goggles to avoid injury during dissections. •

☐ 1  Trim away any excess adipose tissue, leaving the stub of the optic nerve and extrinsic eye muscles intact.
☐ 2  Locate these structures on the external aspect:
  ☐ **Optic nerve**—The nerve bundle projecting from the posterior of the eyeball
  ☐ **Extrinsic eye muscles** (six)—These may have been cut from your specimen during preparation
  ☐ **Sclera**—The white portion of the vascular layer
  ☐ **Cornea**—The clear, anterior portion of the vascular layer
  ☐ **Iris**—The pigmented region under the cornea
  ☐ **Pupil**—The iris's hole (it may be oblong in your specimen rather than round as in the human)
  ☐ **Conjunctiva**—The thin mucous membrane over the anterior portion of the eye, extending to line the inner eyelid

  ☐ Any other accessory structures that may be present in your specimen
☐ 3  Puncture the sclera with the tip of your scissors (or a scalpel) about 1 cm posterior to the edge of the cornea. Use the scissors to cut a circle around the eye, staying 1 cm from the cornea's margin. Pull the anterior portion away from the posterior portion.
☐ 4  Identify these features:
  ☐ **Vitreous body**—The thick jellylike substance of the posterior cavity
  ☐ **Aqueous humor**—A watery liquid in the anterior cavity
  ☐ **Lens**—A plastic-like ball of translucent tissue (in fresh specimens it is very clear; hold it up and try to look through it)
  ☐ **Ciliary body**—A ring of ridges around the outside of the iris's margin (the ridges are formed by the ciliary muscles within)
  ☐ **Retina**—A thin film of gray matter loosely associated with the inside posterior wall of the eyeball (it may have fallen away, appearing as a crumpled mass still attached at the **optic disk**)
  ☐ **Choroid**—The pigmented region of the vascular layer in the posterior region of the eyeball's wall, deep to the sclera, superficial to the retina. In some mammals, but not in humans, an iridescent **tapetum lucidum** can be seen in the choroid.

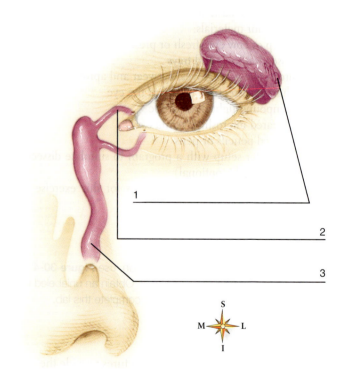

**FIGURE 30-1** Label these parts of the lacrimal apparatus on the lines provided and on the blanks in the Lab Report at the end of this exercise.

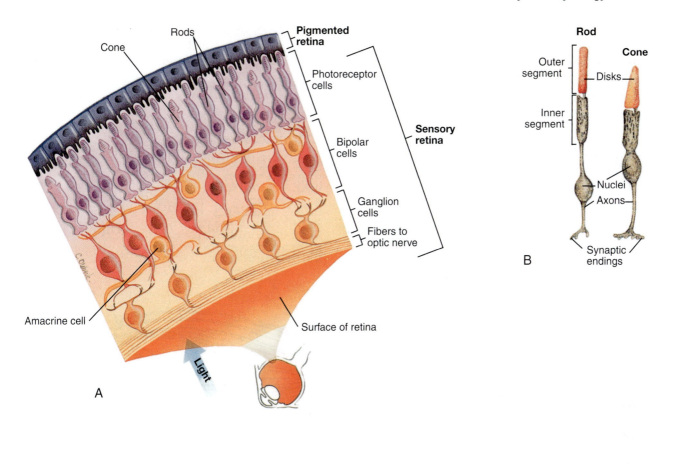

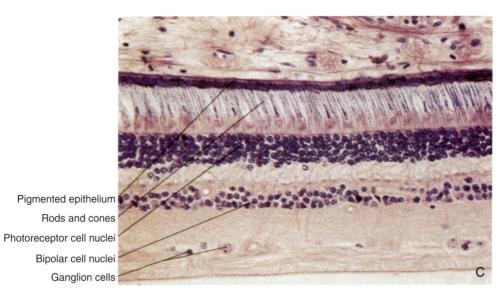

**FIGURE 30-2** Cell layers of the retina. **A,** Pigmented and sensor layers of the retina. **B,** Rod and cone cells. Note their variation in the general structure of a neuron. **C,** Photomicrograph of the retina.

*Coloring Exercises:* **Eye**

Use colored pens or pencils to shade in both the figure and the labels. Each red numeral in the figure corresponds to a matching red numeral following the appropriate label.

FIBROUS LAYER 1
   CORNEA 2
   SCLERA 3
VASCULAR LAYER 4
   CHOROID 5
   CILIARY BODY 6
   IRIS 7
   LENS 8
INNER LAYER 9
   RETINA 10
   AQUEOUS HUMOR 11
   VITREOUS BODY 12
   CONJUNCTIVA 13

**Extrinsic Muscles**
SUPERIOR RECTUS 14
INFERIOR RECTUS 15
LATERAL RECTUS 16
SUPERIOR OBLIQUE 17
INFERIOR OBLIQUE 18

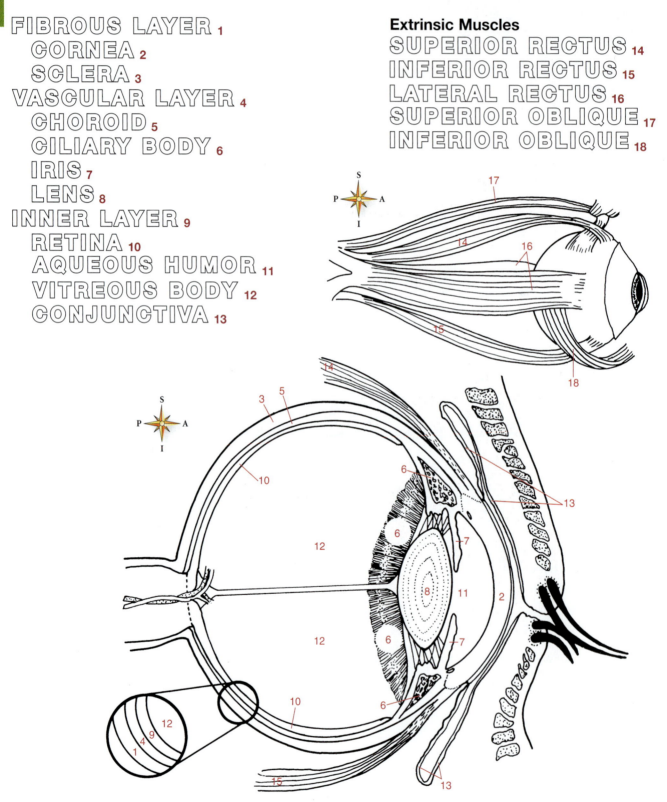

**FIGURE 30-3** Structures of the eye.

LAB EXER

# Visua

Human vision is a c
**reception** by photor
to appropriate brain
mind. The activitie
demonstrate just a fe
human sense of visio

## BEFORE YO

☐ Read the approp
☐ Set your learnir
    should be able to
    ☐ demonstrate
        demonstratio
☐ Prepare your ma
    ☐ meter stick
    ☐ wall chart: as
    ☐ wall chart: Sı
    ☐ pack of assort
    ☐ color blindne
☐ Read the directic
    procedure.

*Study* **TI**

Develop a concep
the pathway of ligh
might want to take
pathway of light th
or the *A&P Surviva*

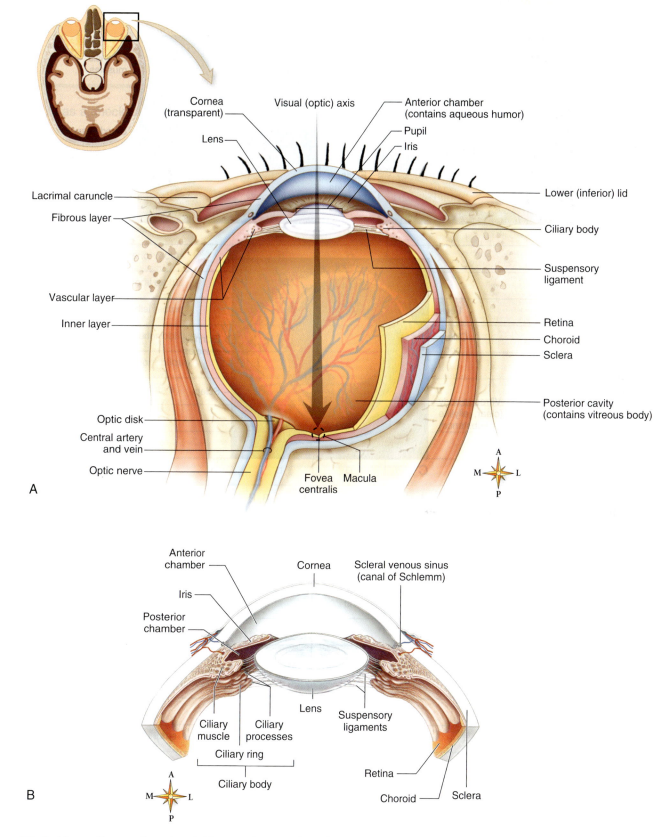

**FIGURE 30-4** Mammalian eye dissection. **A,** Horizontal section through the eyeball, superior view. **B,** Lens, cornea, iris, and ciliary body. Note the suspensory ligaments that attach the lens to the ciliary body.

**Figure 30-**

1. ____
2. ____
3. ____

**Fill-in**

1. ____
2. ____
3. ____
4. ____
5. ____
6. ____

**Put in Ord**

1. ____
2. ____
3. ____
4. ____
5. ____

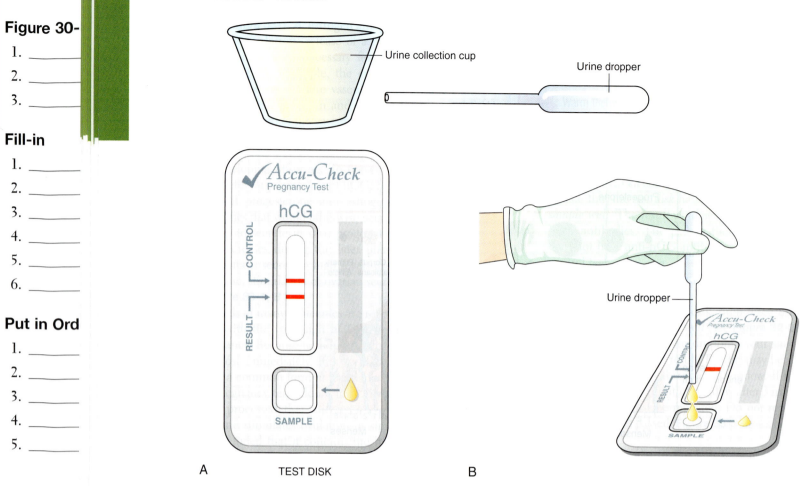

A                    TEST DISK                    B

**FIGURE 33-2**  **A,** Pregnancy test kit. **B,** Apply three drops of urine to the sample well with the urine dropper.

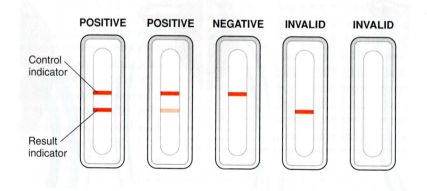

**FIGURE 33-3**  Interpretation of pregnancy test results. A line of *any* shade (light or dark) in the result window is a positive result. No line in the result window is a negative result. However, if there is no red line in the control window (regardless of the presence or absence of a line in the result window), then the test did not work and the result is invalid.

Name: ______________________________ Date: __________ Section: ________

# LAB REPORT 33
# Hormones

Use your textbook or class notes to fill in the following tables.

| HORMONE | GLAND (SOURCE) | TARGET | ACTION |
|---|---|---|---|
| Antidiuretic hormone | | | |
| Oxytocin | | | |
| Growth hormone | | | |
| Thyroid-stimulating hormone | | | |
| Adrenocorticotropic hormone | | | |
| Luteinizing hormone | | | |
| Follicle-stimulating hormone | | | |
| Prolactin | | | |
| Thyroid hormones | | | |
| Calcitonin | | | |
| Parathyroid hormones | | | |
| Epinephrine | | | |
| Aldosterone | | | |
| Cortisol | | | |
| Melatonin | | | |
| Thymosin | | | |
| Insulin | | | |
| Glucagon | | | |
| Testosterone | | | |
| Estrogen | | | |
| Progesterone | | | |
| Prostaglandin(s) | | | |

| TEST | SAMPLE | RESULT PAD | CONTROL PAD | RESULT | EXPLANATION |
|---|---|---|---|---|---|
| Pregnancy (hCG) | Pregnant | | | hCG<br>present (+)<br>not present (−) | |
| | Nonpregnant | | | hCG<br>present (+)<br>not present (−) | |

## Discussion Questions (answer these questions on a separate sheet of paper)

1. Explain how hCG secretion is regulated. Is it secreted by a pregnant woman or her offspring?
2. hCG depresses some reactions of the immune system. What adaptive advantage do you think this has?

# Blood

Blood is the fluid tissue that circulates within the cardiovascular system. Typical of connective tissues, it has a dominant extracellular matrix. In this case the matrix is a liquid called **plasma.** Suspended in the plasma are blood cells or **formed elements. Figure 34-1** shows the major components of human blood. The characteristics of a person's blood vary according to age, sex, metabolic condition, health condition, genetics, and other factors. Clinical evaluation of blood characteristics is important in assessing the condition of patients. This exercise provides demonstrations of just a few of the more common tests. Many blood tests are done by automated machines, at least in some areas, but it is important to understand the basis of these tests to interpret them correctly.

## BEFORE YOU BEGIN

☐ Read the appropriate chapter in your textbook.
☐ Set your learning goals. When you finish this exercise, you should be able to:
  ☐ name the components of human blood tissue and identify their functions
  ☐ demonstrate how common blood tests are performed
  ☐ interpret the meaning of important blood tests
  ☐ state the limitations of clinical blood testing
☐ Prepare your materials:
  ☐ blood sample (see *safety first* box) **Note:** You will be using artificial blood samples purchased for the purpose of performing these exercises. No actual blood will be used.
  ☐ paper towels, lab wipes
  ☐ clean microscope slides
  ☐ bibulous paper
  ☐ distilled water (in dropper bottle)
  ☐ Wright stain (in dropper bottle)
  ☐ wax pencil
  ☐ hemocytometer (glass)
  ☐ BMP LeukoChek white blood cell (WBC) count kit
  ☐ heparinized capillary tubes
  ☐ microhematocrit centrifuge
  ☐ capillary tube sealing clay or plugs
  ☐ metric ruler (mm)
  ☐ Tallquist test paper and hemoglobin (Hb) scale (or hemoglobinometer kit)
  ☐ toothpicks
  ☐ blood typing sera: *anti-A, anti-B, anti-D*
  ☐ slide warming box
  ☐ compound light microscope
  ☐ protective gear: gloves, safety eyewear, and apron
☐ Read the directions and safety tips for this exercise *carefully* before starting any procedure.

## safety first

This exercise calls for the use of synthetic human blood samples. These tests should be performed as a demonstration by the instructor using simulated human blood designed specifically for student demonstrations. Follow current laboratory practice by wearing a lab coat, gloves, and protective eyewear while in the lab. •

## A. SAMPLING BLOOD

Blood sampling for clinical testing usually requires only tiny amounts of blood, often only a few drops. The usual method for collecting small blood samples is to draw it from a punctured finger. This activity tells you how blood can be drawn for clinical testing. **Please remember: You will not be collecting any blood from yourself or a classmate. You will be using only artificial blood supplied by your instructor.**

## B. BLOOD SMEAR

Investigate blood tissue by looking at several prepared blood smear slides provided by your instructor. As the name implies, it is merely some blood smeared on a slide, then prepared for microscopic examination.

☐ 1 Scan prepared blood slides under low power on your microscope until you find an area where the cells are thinly spread. Switch to high power and observe the blood cells on the slide. Do all the cells look the same? Are different cells (if there are any) distributed equally? What unique characteristics do these cells have (compared with other cells you have observed)? Describe the pattern or order of the cells in the tissue observed. Write your answers to these questions, and any other observations, in the Lab Report at the end of this exercise.

**Hint** If your microscope has an oil-immersion objective, you may want to use it to see the blood cells more clearly. After focusing on high-dry, rotate the nosepiece *halfway* to the high-oil objective. Put a drop of immersion oil on the slide, then rotate the high-oil objective into the oil drop. You may need to increase the lighting. •

## C. DIFFERENTIAL WHITE BLOOD CELL COUNT

The larger nucleated cells that you may have observed in the previous activity are **WBCs,** or **leukocytes.** WBCs are far less numerous than **RBCs (erythrocytes)** and **platelets (thrombocytes). Figure 34-2**

## Coloring Exercises: Human Blood

Use colored pens or pencils to shade in both the figure and the labels. Each red numeral in the figure corresponds to a matching red numeral following the appropriate label.

WHOLE BLOOD 1
PLASMA 2
PROTEINS 3
ALBUMINS 4
GLOBULINS 5
FIBRINOGEN 6
WATER 7
OTHER SOLUTES 8
IONS 9
NUTRIENTS 10
WASTE PRODUCTS 11
GASES 12

REGULATORY SUBSTANCES 13
FORMED ELEMENTS 14
THROMBOCYTES 15
LEUKOCYTES 16
NEUTROPHILS 17
LYMPHOCYTES 18
MONOCYTES 19
EOSINOPHILS 20
BASOPHILS 21
ERYTHROCYTES 22

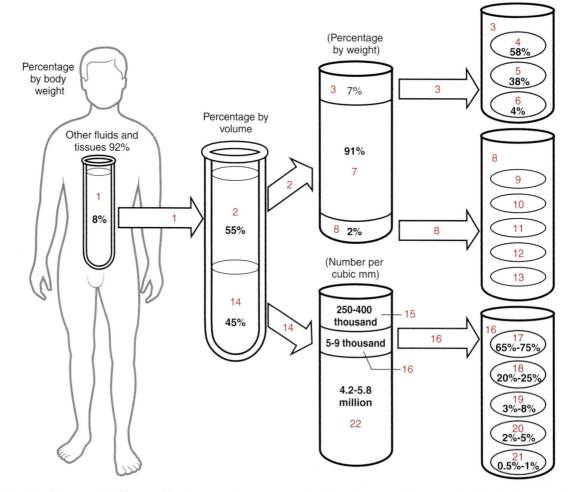

**FIGURE 34-1** Components of human blood tissue. (*Note:* Do not color inside the numerals or % symbols in the figure so that you can see them more easily when you study them later.)

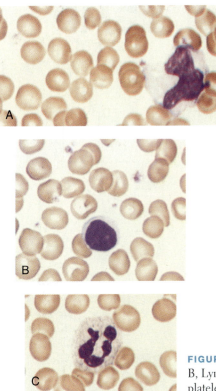

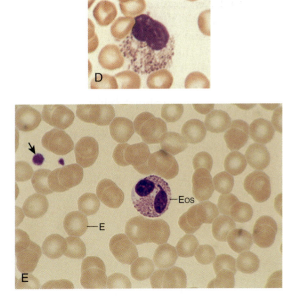

**FIGURE 34-2** Representative human blood cells. A, Monocyte. B, Lymphocyte. C, Neutrophil. D, Basophil. E, Erythrocyte, eosinophil, platelet. *E,* Erythrocyte; *Eos,* eosinophil; *arrow,* platelet.

shows the different types of blood cells. WBCs have numerous protective functions in the body, so their numbers sometimes change in response to changes in health.

Different leukocytes have different functions, so the changes may be reflected in changes in the relative proportions of different types of leukocytes. Clinically, information about relative proportions of leukocyte types is very useful. A test that gives this information is the **differential WBC count.**

☐ 1  Focus on a portion of the blood smear, as in the previous activity. Look around the field for any leukocytes.

☐ 2  In the Lab Report at the end of this exercise, mark off how many leukocytes you see of each type until you reach 100 total cells.

Obviously, you'll have to move around to different fields to reach 100 cells. Move the slide in the manner shown in **Figure 34-3** so that you do not count the same portion of the slide twice.

☐ 3  Determine the total number of WBCs of each type. If the total for all types is 100, the number for each type is also the percent of that type (i.e., because 100 is the total, 1 cell is also 1% of the cells).

---

📌 *landmark characteristics*

Below are the normal ranges for the WBC types. Also included in parentheses is one reason that the percentage of each WBC type could be higher than normal.

- **Monocytes:** 3% to 8% (chronic infections)
- **Lymphocytes:** 20% to 25% (antibody reactions)
- **Neutrophils:** 65% to 75% (acute infections)
- **Eosinophils:** 2% to 5% (allergic reactions)
- **Basophils:** 0.5% to 1% (inflammatory condition)

---

📌 *landmark characteristics*

**WBCs** have nuclei and are distinctly larger than **RBCs. Agranulocytes** *(agranular leukocytes)* have no granules (spots) visible in the cytoplasm, but **granulocytes** *(granular leukocytes)* do (see **Figure 34-2**).

- **Monocytes** are very large, agranular leukocytes with large, variably shaped nuclei.
- **Lymphocytes** are agranular WBCs that are almost as small as RBCs. Lymphocyte nuclei are sometimes so large that a lymphocyte appears to have no cytoplasm.
- **Neutrophils** are granular WBCs with small, pinkish granules and lobed nuclei that resemble several links of sausage.
- **Eosinophils** have red granules and two-lobed, dark nuclei.
- **Basophils** have fewer granules, which are a bluish tint and of variable size. Basophils have large, two-lobed or kidney-shaped nuclei.

☐ 4  How do your results compare with the normal values given in the box?

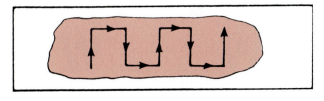

**FIGURE 34-3** Move the slide in this manner to prevent counting the same cells more than once.

# D. WHITE BLOOD CELL COUNT

**WBC counts** are clinical tests that estimate the number of leukocytes present in each cubic millimeter ($mm^3$) of blood. A known volume of a solution called the *diluent* is used to dilute a blood sample (also of a known volume) so that the cells are more spread out and therefore easier to count. The diluted sample is placed on a special glass microscope slide called a **hemocytometer.** The hemocytometer, shown in **Figures 34-4** and **34-5,** has two counting grids etched on it. When filled correctly, just the right volume is present under the coverslip and over the counting grid.

The instructions that follow are for the BMP LeukoChek test **(Figure 34-6).** The BMP LeukoChek test is for the microscopic counting of leukocytes in whole blood. If you are using another method,

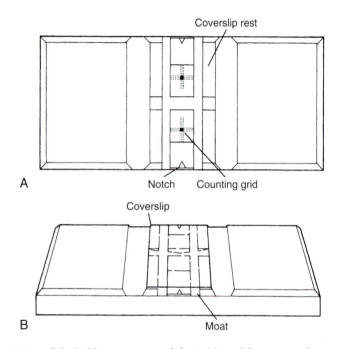

**FIGURE 34-4**  Hemocytometer slide. A, Viewed from top, without coverslip. B, Viewed from front, with coverslip.

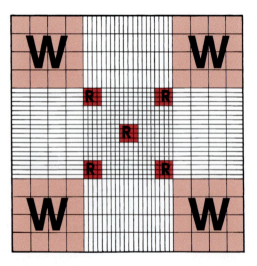

**FIGURE 34-5**  Detail of hemocytometer counting grid. Areas used for counting WBCs are marked *W*, and the areas used for counting RBCs are marked *R*.

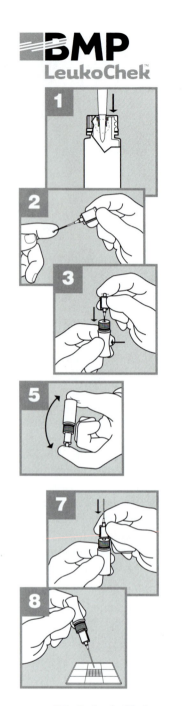

**FIGURE 34-6**  LeukoChek test.

follow the directions supplied by the manufacturer of the test kit or by your instructor. In any case, be sure to heed all safety advice.

> **Hint** ▶ Artificial whole blood (20 μl) will be added to the BMP Leuko-Chek reservoir from a well-mixed anticoagulated tube of whole blood. The manufacturer provides a capillary tube that fills via capillary action to exactly 20 μl of artificial whole blood. Each reservoir contains 1.98 mL of a 1% buffered ammonium oxalate solution. This provides for a ratio of 1:100 sample of total volume.  •

□ 1 Puncture the cap diaphragm with the protective shield on the pipette assembly.

□ 2 Remove the shield and fill the tube with artificial whole blood from a sample provided by your instructor that supplies you with the artificial whole blood. Be sure that the capillary tube fills completely. When the blood reaches the end of the capillary tube, it will stop automatically, filling the tube with 20 µL of artificial whole blood. Gently wipe excess blood from the outside of the capillary tube.

□ 3 Squeeze the reservoir slightly to expel a volume of air. Maintain pressure on the reservoir while inserting the pipette with whole blood into the reservoir. Be sure to simultaneously cover the top opening of the capillary tube holder.

□ 4 Release pressure from the reservoir and then from the capillary tube opening. This will cause the artificial whole blood to be drawn into the diluent.

□ 5 Gently squeeze the reservoir to rinse the capillary tube, taking care not to expel any mixture of diluent and whole blood through the top of the opening of the inserted capillary tube holder. Place your finger over the top of the opening of the capillary tube holder and gently invert the reservoir several times to ensure proper mixing.

□ 6 Wait 10 minutes before attempting to count leukocytes.

□ 7 Remove the pipette assembly, and invert and seat the assembly in a reverse position into the top of the cap. This changes the pipette assembly from a collection device to a dropper.

□ 8 Discard the first 3 or 4 drops, and then expel the reservoir solution into a hemocytometer with a Neubauer grid for counting the leukocytes.

□ 9 A leukocyte count is performed under 100× magnification. Leukocytes are counted in all nine large squares of the counting chamber. Add 10% to the total number and then multiply this value by 100 to get the total leukocyte count. For example, if 60 cells are counted, the total count would be 60 + 6 × 100, which would represent a total count of 6600 leukocytes per cubic millimeter.

### landmark characteristics

Normal: 4500 to 11,000 WBCs/mm³
↑ in acute infections, trauma, some cancers
↓ in some forms of anemia or during chemotherapy

## E. HEMATOCRIT

Blood is composed of a number of components. Plasma is the extracellular fluid matrix of blood tissue. Mostly water, it also contains ions (e.g., $Na^+$ and $Cl^-$), **plasma proteins** (e.g., *antibodies* and *albumin*), and other dissolved substances (e.g., *urea*, $O_2$, and *glucose*). The formed elements (blood cells or *blood solids*) include

RBCs, WBCs, and platelets. You have already completed blood tests that look at numbers and proportions of WBCs and total numbers of RBCs. The test demonstrated in this activity explores the relative proportions of blood solids and blood plasma.

The **hematocrit** test is also called the **packed cell volume (PCV)** determination because it involves the packing of all the cells in a blood sample at one end of a tube. The sample is placed into a *centrifuge*, a machine that spins a tube of blood and allows centrifugal force to push the heavy blood cells to the bottom. Once the cells are packed, the proportion of cells to total volume can be calculated. The results are used to determine whether there is a shortage of RBCs (as in some *anemias*) or an excess of RBCs (*polycythemia*), or dehydration.

□ 1 Fill a heparinized capillary tube about 75% full of artificial blood by touching it to the sample whole blood provided by your instructor. Heparinized tubes are red-tipped for easy identification.

□ 2 Seal the unused end of the tube by holding it near the tip and pushing it into a tray of sealing clay at a 90-degree angle with a twisting motion. A plug of clay will then seal the tube. Remove the tube. Plastic plugs are also available for this purpose.

### ! safety first

Capillary tubes are easily broken, causing injury and contamination by the blood. Have the instructor demonstrate the proper technique for sealing them before attempting it yourself. Handle capillary tubes with gloved hands. •

□ 3 Place the tube in one of the numbered grooves of a microhematocrit centrifuge with the sealed end facing the perimeter of the centrifuge tray. Close the top of the centrifuge according to the manufacturer's instructions and spin the tubes for 4 minutes.

□ 4 Remove your tube from the centrifuge and notice that the RBCs are now packed in one end of the tube. There may be a thin, lightly colored, *buffy coat* topping the RBCs. This coat is actually the WBCs and platelets, which are slightly less dense than RBCs. The rest of the tube contains the yellowish, transparent plasma.

□ 5 Lay the tube on a paper towel next to a metric ruler and measure the total length of the sample, from the clay-RBC border to the end of the plasma. Next, measure the RBC portion only. Record your results here:
□ _____ mm Total length
□ _____ mm Packed cell length

□ 6 Divide the length of the RBC section by the total sample length. Multiply your answer by 100 to arrive at the percent volume of packed RBCs. This is the hematocrit value, reported with or without the percent symbol. Report your results in the Lab Report at the end of this exercise. Use the normal values given in the Landmark Characteristics box to determine whether your results fall within the normal range.

## F. HEMOGLOBIN DETERMINATION

**Hemoglobin (Hb)** is the pigment inside RBCs that has an affinity for $O_2$ (and $CO_2$), so it functions to transport blood gases. Although knowing the number or proportion of RBCs in the blood may hint at the total Hb content, a *hemoglobin determination* gives a relatively accurate figure. Most methods operate on the principle that the denser the color of the sample, the higher the content of Hb pigment. Of course, the Hb has to first be liberated from the RBCs and allowed to diffuse evenly throughout the plasma. The more accurate methods use *colorimeters* or *spectrophotometers* that measure the percentage of filtered light transmitted through the sample. Some clinics use a handheld version of this type of instrument called a *hemoglobinometer*. A less accurate method, the *Tallquist method*, is used here to demonstrate Hb determination. (If you have a hemoglobinometer and wish to use it, follow the directions that follow those for the Tallquist method.)

**Tallquist Method:**

☐ 1 Place a drop of blood supplied by your instructor on the special paper in the Tallquist test kit. The RBCs break open in the paper.

☐ 2 Allow the drop to dry long enough to lose its glossy appearance but not long enough for it to become brown.

☐ 3 Compare the color of the blood spot to the colors in the Tallquist chart in your kit. Determine the Hb content by matching colors as closely as possible. Record your result in grams of Hb per 100 mL of blood (g/100 mL) in the Lab Report at the end of this exercise. How does it compare with normal values?

**Hemoglobinometer Method:**

☐ 1 Obtain a hemoglobinometer and locate the eyepiece, sample chamber, cover plate, and clip. Also obtain some hemolysis applicators.

☐ 2 Let a drop of sample blood fall on the open sample chamber.

☐ 3 Hemolyze (break open) the blood cells in the sample chamber by agitating gently with a hemolysis applicator. When hemolysis is complete (after about 45 seconds), the blood will have changed to a clear red solution.

☐ 4 Clip the cover on the sample chamber and slide the whole assembly into the slot on the side of the hemoglobinometer.

☐ 5 Press the illuminating button on the bottom of the hemoglobinometer as you look into the eyepiece. Move the sliding knob on the side of the hemoglobinometer until both sides of the split green field match in intensity.

☐ 6 Read the hemoglobin concentration directly from the scale and record your result in the Lab Report at the end of this exercise.

## G. BLOOD TYPING

All cells have different proteins on the surface of their cell membranes that act as identification tags. The human immune system has cells and chemicals that can recognize proteins as *nonself* proteins. In this way, immune processes can try to destroy or inhibit proteins (and cells to which they may be attached) that are foreign to a person's body. The identifying proteins on cell surfaces are determined by heredity. If a foreign tissue is transplanted into someone's body, as in blood transfusions, the immune system will destroy new blood cells that don't have the *self* type of marker proteins.

To prevent this *tissue rejection* from happening, biologists have determined which blood cell marker proteins elicit life-threatening immune reactions. Biologists have also devised tests to determine the presence of these proteins and systems of naming them. This demonstration shows you how some of these **blood typing** tests work.

Before beginning the demonstration, you must learn some new terms. The term **antigen** refers to a molecule that elicits an immune response. In the case of blood transfusions, marker proteins on the donor's RBCs can be antigens (if they are nonself relative to the recipient). The term **antibody** refers to a type of plasma protein produced by the immune system. Antibodies react with non-self markers (antigens) and try to destroy or inhibit them. One type of antigen-antibody reaction is the **agglutination reaction,** in which antigens are "glued" together with antibodies like flies on flypaper. This clumping (agglutination) is the reaction commonly seen when incompatible blood types are transfused.

In the **ABO system** of blood typing, two blood antigens are important: the **A antigen** and the **B antigen.** People with *type A* blood have the A antigen but not the B antigen. *Type B* blood has the B antigen but not the A antigen. *Type AB* blood has both A and B antigens, whereas *type O* blood has neither A nor B antigens. To type blood in the ABO system, blood is mixed with *anti-A serum* and *anti-B serum.* Serum is plasma with the clotting factors removed. Anti-A serum contains antibodies that cause agglutination when the A antigen is present. Anti-B serum contains antibodies that react with the B antigen.

**safety first**

Handle artificial blood serum with the same caution that you would use when handling any blood sample.  •

☐ 1  Place a clean microscope slide on a paper towel and write "Anti-A" near the left side and "Anti-B" near the right side. Use a wax pencil to make two 1-cm circles on the slide, one on the right and one on the left.

☐ 2  Put a drop of artificial blood supplied from the samples provided by your instructor in each circle.

☐ 3  To the drop of blood in the left circle, add a drop of anti-A serum *without touching the dropper to the blood*. Similarly, add a drop of anti-B serum to the right blood drop. Using a different toothpick for each circle, mix the serum and the blood.

☐ 4  Check each mixture for a clumping (agglutination) reaction, in which the mixture changes from a uniform reddish color to distinct dark clumps in a transparent medium. If a reaction occurred in the left circle (anti-A), then the A antigen is present. If no reaction occurred, then the antigen is not present. According to the ABO system, what type is the blood (i.e., A, B, AB, or O)? Check your results against **Figure 34-7**.

Another commonly used blood typing system is the **Rh system**, which deals with the **Rh antigen.** Sometimes called the *D antigen*, this blood cell marker does not elicit an immune reaction in someone who does not have it until a second exposure to the antigen because the immune system doesn't make **anti-D antibodies** until it has been exposed to the D antigen. Once exposed, however, an army of antibodies (and anti-D–producing cells) stands ready to react with the D antigen. A person with the antigen is said to be *Rh-positive* and a person without the antigen is termed *Rh-negative*. The Rh typing test is similar to the ABO test:

☐ 1  Draw a circle on a clean glass slide and place a drop of blood within it (as described earlier).

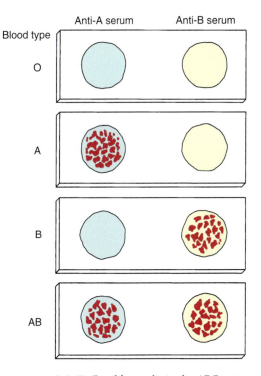

**FIGURE 34-7**  Possible results in the ABO test.

☐ 2  Add a drop of anti-D serum and place the slide on a slide warmer. This device warms the slide to about 45° C and tilts it back and forth. Because the Rh reaction is less severe than reactions in the ABO system, such treatment is necessary to help the reaction so that it can be easily observed. Look for fine clumping. What Rh type is the sample?

Name: _____    Date: _____ Section: _____

## LAB REPORT 34
# Blood

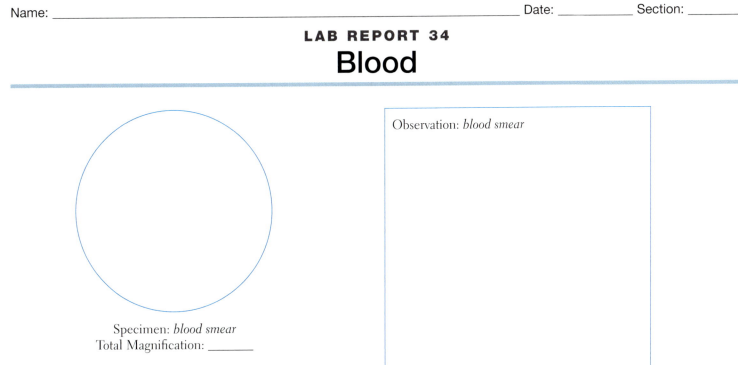

Specimen: *blood smear*
Total Magnification: _____

Observation: *blood smear*

## Differential WBC Count Tally Grid

| TYPE | MONOCYTE | LYMPHOCYTE | NEUTROPHIL | EOSINOPHIL | BASOPHIL |
|------|----------|------------|------------|------------|----------|
| Tally marks | | | | | |
| Totals | | | | | |
| Percentages | | | | | |
| Normal value | | | | | |
| Interpretation | | | | | |

## Blood Test Results

| SUBJECT: | SEX: ☐ MALE   ☐ FEMALE | | |
|----------|----------|----------|----------|
| TEST | NORMAL VALUE | RESULT | INTERPRETATION |
| WBC count | | | |
| RBC count | | | |
| Hematocrit | | | |
| Hemoglobin | | | |
| Method: _____ | | | |
| ABO type | | | |

## Put in Order

1. _____
2. _____
3. _____
4. _____
5. _____
6. _____
7. _____
8. _____
9. _____
10. _____
11. _____
12. _____
13. _____
14. _____

## Fill-in

1. _____
2. _____
3. _____
4. _____
5. _____
6. _____
7. _____
8. _____
9. _____
10. _____
11. _____
12. _____
13. _____
14. _____
15. _____

## Put in Order (rearrange these structures in the order in which blood passes through them; assume that the blood is about to leave the right atrium)

aorta
aortic semilunar valve
left ventricle
left atrium
lung capillaries
mitral valve
pulmonary semilunar valve
pulmonary arteries
right ventricle
superior/inferior vena cava
tissues of the body
tricuspid valve
pulmonary trunk
pulmonary veins

## Fill-in (complete each statement with the correct term)

1. The flaplike lateral wall of each atrium is called the _?_.
2. The _?_ valve is also known as the mitral valve or left AV valve.
3. The right AV valve is also known as the _?_ valve.
4. The aortic semilunar valve has _?_, pocketlike flaps of tissue.
5. The _?_ are fibrous structures that prevent the cuspid valves from prolapsing (bending backward).
6. One-way flow of blood from the right ventricle is ensured by the presence of the _?_ valve.
7. Mitral valve prolapse, which is abnormal, may allow blood to enter the _?_ during contraction of the left ventricle.
8. The small cardiac vein and right coronary artery can be found along the right _?_ sulcus.
9. The anterior interventricular branches of the left coronary artery and cardiac vein are found along the anterior _?_ sulcus.
10. The _?_ is a muscular wall between the left and right ventricles.
11. The myocardium of the _?_ ventricle is thicker than the other ventricle.
12. The wall of the aorta is _?_ (thicker/thinner) than the wall of the superior vena cava.
13. The _?_ are beamlike processes of the inner face of the myocardium.
14. The "point" of the heart is called the _?_.
15. In the sheep heart, the right atrium is _?_ to the right ventricle.

# LAB EXERCISE 36
# Electrical Activity of the Heart

In the previous exercise, you learned some of the major anatomical features of the heart. This exercise challenges you to take a step beyond the basics and learn about the electrical nature of the heart. As part of your study, you will demonstrate one of the most commonly used clinical tools for assessing the health of the heart: electrocardiography (ECG).

This exercise calls for any of several electrocardiographs commonly available in educational laboratories (Activities A and B). A computer-based approach is presented as an alternate method in Activities C and D.

## BEFORE YOU BEGIN

☐ Read the appropriate chapter in your textbook.
☐ Set your learning goals. When you finish this exercise, you should be able to:
  ☐ describe the conduction system of the heart
  ☐ demonstrate electrocardiography
  ☐ identify the features of a typical electrocardiogram (ECG) and explain their significance
☐ Prepare your materials:
  ☐ electrocardiograph apparatus or computerized ECG program
  ☐ alcohol swabs
  ☐ electrocardiographic electrodes, electrode tape, and gel
  ☐ table or cot
  ☐ metric (mm) ruler
☐ Read the directions and safety tips for this exercise *carefully* before starting any procedure.

---

**⓵ safety first**

The apparatus used in this exercise is powered by electricity, so appropriate cautions should be taken to avoid electrical hazards. Students with known heart problems may volunteer as subjects in the study of resting ECGs but should not participate in activities involving exercise or other stressful situations.  •

---

**Study TIPS**

Reading and interpreting an ECG requires a lot of practice and expertise. However, there are two variations from a normal ECG that may indicate an abnormality: variation in wave height (elevated or depressed) or variation in normal time intervals between waves. Refer to the Cardiac Dysrhythmia section in this lab exercise. Your textbook also provides examples of ECG strip recordings.

## A. ELECTROCARDIOGRAPHY

The **cardiac cycle** is the pattern of physiological events exhibited during each beat of the heart. One approach is to describe the cardiac cycle as atrial, then ventricular, **systole** (contraction), and **diastole** (relaxation). In other words, the atria contract and relax, and the ventricles contract and relax, producing one pumping cycle of the heart. During the cardiac cycle, a variety of physiological characteristics of the heart and pumped blood change in a way that is measurable. For example, the **blood pressure** within each chamber changes with the volume of the chamber and the strength of contraction of the myocardium. Certain **heart sounds** are heard as the valves snap shut and portions of the myocardium contract. The electrical properties of the heart change as action potentials travel through the myocardium. This electrical aspect of the cardiac cycle is of interest in this activity.

As you know, all muscle requires *excitation* of the cell membrane coupled with contraction involving filaments within the cell. This excitation, taking the form of action potentials, is an electrical event in which the polarity of electrical potential across the cell membranes changes. Because so much myocardial tissue depolarizes at more or less the same time, a great deal of current flows around the heart tissue, and the depolarization is easily detected from the outside of the body. Around the beginning of the twentieth century, a Dutch scientist named Einthoven was one of the first to use a recording voltmeter to measure this electrical activity of the heart and to describe the changes found in a typical cardiac cycle. This technique is called **electrocardiography**.

The **ECG** is a waveform graph produced by a recording voltmeter called an **electrocardiograph**. The graph is a line on a paper or monitor that rises and falls with changes in voltage (electric potential) between two points. **Figure 36-1, A**, shows the voltage changes seen in one typical cardiac cycle. For convenience, each wave is given a name:

☐ **P wave**—This wave represents depolarization of the atrial walls, which precedes atrial systole.
☐ **QRS complex**—This set of waves results from the more or less concurrent ventricular depolarization (preceding ventricular systole) and atrial repolarization (signaling the onset of atrial diastole).
☐ **T wave**—This wave represents ventricular repolarization, which precedes ventricular diastole.

The length of time during waves and between waves indicates the efficiency of conduction through the myocardium. Lengthened ECG portions are often evidence of blocked conduction. To understand the significance of conduction blockage, we must first outline the structure and function of the conduction system:

☐ **Sinoatrial (SA) node**—This specialized section of cardiac muscle in the upper lateral wall of the right atrium generates its

☐ **5** Using lead II results, measure these values:

☐ Heart rate (in beats per minute)—Remember that the paper rate is 25 mm/sec (each millimeter = 0.04 seconds). **Tachycardia** is the condition of a heart rate greater than 100 beats/min and may be caused by excitement (stress), high body temperature, or toxicity (see **Figure 36-5, D**). **Bradycardia** is the condition of lowered heart rate (less than 60 beats/min). It is normal in conditioned athletes because they have a high stroke volume, or it may result from excessive vagal stimulation (see **Figure 36-5, C**).

☐ PR (PQ) interval—If greater than 0.2 seconds, a **first-degree heart block** could be present (see **Figure 36-5, B**). This may result from inflammation of the AV bundle, which slows down conduction of a signal to the ventricles. If the PR interval is 0.25 to 0.45 seconds (and some P waves are not followed by QRS complexes), **second-degree heart block** may be present. This could result from AV node damage or excessive vagal stimulation. If the P wave seems independent of the QRS complex, with a P rate of about 100 per minute and a QRS of less than 40 per minute, **complete heart block** may be present.

☐ Your instructor may ask you to make other measurements and consult your textbook or a reference book for guidance in interpreting your results.

## B. THE AUGMENTED LEADS

Many computerized ECG programs are designed to print out results of the standard (appendicular) leads, as well as three **augmented leads.** You may recall that the Einthoven triangle diagrammatically represents the direction of electric potential measurements using the three standard electrode arrangements (leads I, II, and III; **Figure 36-4, A**). Einthoven described a number of mathematical relationships among data produced by the three leads. For example, knowing data from any two leads is sufficient to mathematically reconstruct a data line for the third lead. Information from combinations of two leads can be manipulated to produce the three augmented, or attenuated, leads. These are not bipolar leads or electrode placements but are the result of combining two negative electrodes and measuring them against one positive electrode. Thus three more leads are produced:

☐ $aV_R$—The LA and LL are combined to form the negative electrode; RA is the positive electrode.

☐ $aV_L$—The RA and LL electrodes form the negative electrode; LA is the positive electrode.

☐ $aV_F$—The RA and LA electrodes form the negative pole; LL is the positive pole.

Using the concept of the Einthoven triangle, the augmented leads are each shifted 90 degrees from the standard leads (see **Figure 36-4, B**).

Current clinical practice calls for recording 12 leads in all. So far, we have discussed the three standard leads and the three

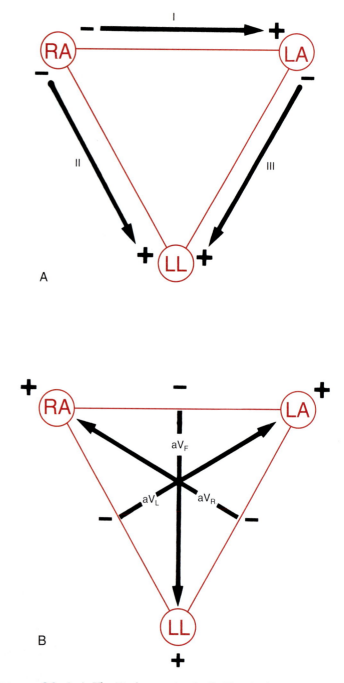

**FIGURE 36-4** **A,** The Einthoven triangle. **B,** The Einthoven triangle superimposed with the three augmented leads.

augmented leads. The remaining six leads are the **chest,** or **precordial,** leads. The chest leads use RA, LA, and LL combined to form the negative pole and each of six chest electrodes to form the positive poles. The chest leads are named $V_1$ through $V_6$.

What is the advantage of chest leads?

# clinical applications: *cardiac dysrhythmia*

The normal rhythmical beating of the heart **(Figure 36-5, *A*)** can be disturbed by conditions (e.g., inflammation of the endocardium [*endocarditis*] or myocardial infarction [*heart attack*]) that damage the heart's conduction system. The term **dysrhythmia** refers to an abnormality of heart rhythm.

One kind of dysrhythmia is called a **heart block.** In *AV node block*, impulses are blocked from getting through to the ventricular myocardium, resulting in the ventricles contracting at a much slower rate than normal. On an ECG, there may be a large interval between the P wave and the R peak of the QRS complex (see **Figure 36-5, *B***). *Complete heart block* occurs when the P waves do not match up at all with the QRS complexes—as in an ECG that shows two or more P waves for every QRS complex. A physician may treat heart block by implanting an *artificial pacemaker* in the heart.

**Bradycardia** is a slow heart rhythm—less than 50 beats/min (see **Figure 36-5, *C***). Slight bradycardia is normal during sleep and in conditioned athletes while they are awake (but at rest). Abnormal bradycardia can result from improper autonomic nervous control of the heart or from a damaged SA node. If the problem is severe,

artificial pacemakers can be used to increase the heart rate by taking the place of the SA node.

**Tachycardia** is a very rapid heart rhythm—more than 100 beats/min (see **Figure 36-5, *D***). Tachycardia is normal during and after exercise and during the stress response. Abnormal tachycardia can result from improper autonomic control of the heart, blood loss or shock, the action of drugs and toxins, fever, and other factors.

**Sinus dysrhythmia** is a variation in heart rate during the breathing cycle. Typically, the rate increases during inspiration and decreases during expiration. The causes of sinus dysrhythmia are not clear. This phenomenon is common in young people and usually does not require treatment.

**Premature contractions,** or *extrasystoles,* are contractions that occur before the next expected contraction in a series of cardiac cycles. For example, *premature atrial contractions (PACs)* may occur shortly after the ventricles contract—seen as early P waves on the ECG (see **Figure 36-4, *E***). Premature atrial contractions often occur with lack of sleep, too much caffeine or nicotine, alcoholism, or heart damage. Ventricular depolarizations that appear earlier than expected are called *premature ventricular*

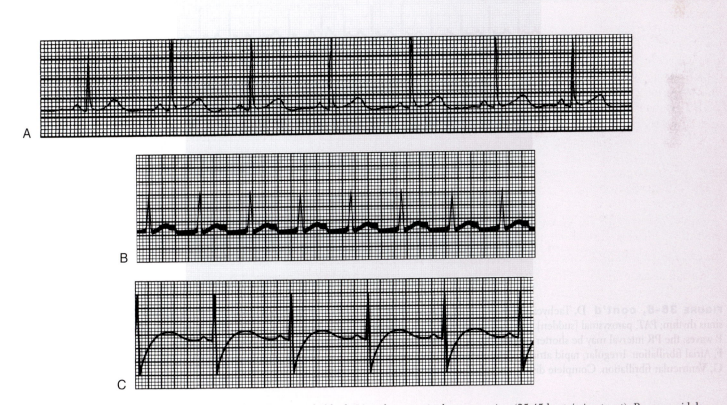

**FIGURE 36-5** **A,** Normal ECG strip chart recording. **B,** AV node block. Very slow ventricular contraction (25-45 beats/min at rest); P waves widely separated from peaks of QRS complexes. **C,** Bradycardia. Slow heart rhythm (less than 60 beats/min); no disruption of normal rhythm pattern.

"120 over 80," or 120 mm Hg/80 mm Hg. Normal ranges of resting blood pressure are shown in **Figure 37-3, B**.

☐ 4 The difference between the systolic pressure and diastolic pressure represents the **pulse pressure** (see **Figure 37-3, A**). As written in an equation:

Pulse pressure = Systolic pressure – Diastolic pressure

For example, if a subject's blood pressure is 120/80, then the pulse pressure would be 40 mm Hg. In other words, 120 mm Hg minus 80 mm Hg equals 40 mm Hg.

☐ 5 The average blood pressure in the arteries is called **mean arterial pressure (MAP)** (see **Figure 37-3, A**). Since diastole (ventricular relaxation period) is about twice as long as systole (ventricular contraction period), the following equation is used to calculate MAP:

$$MAP = \frac{(\text{Diastole pressure} \times 2) + \text{Systolic pressure}}{3}$$

Another way to write the equation for MAP is:

$$MAP = \text{Diastolic pressure} + \frac{\text{Pulse pressure}}{3}$$

For example, if a subject's blood pressure is 120/80, then the MAP would be 93.3 mm Hg. Perform the calculation using one of the equations for MAP and a blood pressure of 120/80 to see if you get the same results.

## C. VARIATIONS IN PULSE AND PRESSURE

Determine the pulse rate and blood pressure in the situations described in the following text, recording the results in the Lab Report at the end of this exercise. Compare the results to each other and to the results for a resting subject. How do you account for the variations?

☐ 1 Test a variety of body positions: lying down (immediately, then after 5 minutes) and standing (immediately, then after 5 minutes).

☐ 2 Test a subject after exercise (immediately after 2-5 minutes of exercise, then once per minute for the next 10 minutes of recovery). Compare your results with those for the postexercise electrocardiogram (Lab Exercise 36—Electrical Activity of the Heart).

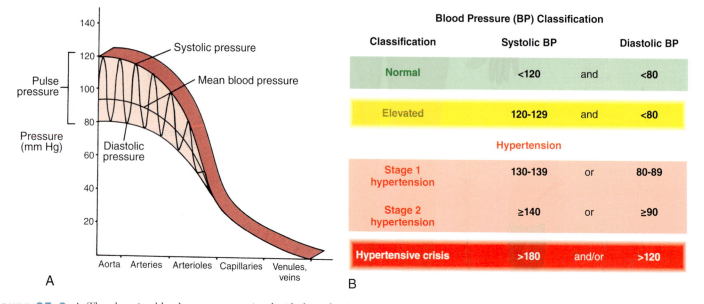

**FIGURE 37-3** **A,** The changing blood pressures associated with the pulse (pressure wave). **B,** Blood pressure ranges. This classification scheme is adapted from the *Joint ACC-AHA 2017 Clinical Practice Guidelines.*

Name: _____ Date: _____ Section: _____

## LAB REPORT 37
# The Pulse and Blood Pressure

| TEST CONDITION | PULSE RATE | PULSE STRENGTH | SYSTOLIC BLOOD PRESSURE | DIASTOLIC BLOOD PRESSURE | PULSE PRESSURE | MEAN ARTERIAL PRESSURE (MAP) |
|---|---|---|---|---|---|---|
| Resting: *radial* | | | | | | |
| Resting: _____ | | | | | | |
| Resting: _____ | | | | | | |
| Lying down: | | | | | | |
| 0 min | | | | | | |
| 5 min | | | | | | |
| Standing: | | | | | | |
| 0 min | | | | | | |
| 5 min | | | | | | |
| After exercise: | | | | | | |
| 0 min | | | | | | |
| 1 min | | | | | | |
| 2 min | | | | | | |
| 3 min | | | | | | |
| 4 min | | | | | | |
| 5 min | | | | | | |
| 6 min | | | | | | |
| 7 min | | | | | | |
| 8 min | | | | | | |
| 9 min | | | | | | |
| 10 min | | | | | | |

In the space provided, write your interpretation of the results reported in the preceding table. For example, tell why the strength of the resting pulse rate may have varied depending on where it was detected. Look for similarities and differences from row to row (how does pulse rate in different body positions compare?). Look for relationships from column to column (does pulse strength relate to pulse pressure?). Use the questions given in the exercise itself for further guidance.

## Interpretation:

# The Circulatory Pathway

Blood pumped from the right ventricle of the heart travels to and through the lungs before returning to the left atrium. This pathway is termed the **pulmonary circulation,** or the *pulmonary loop.* Blood pumped from the left ventricle travels to and through the other organs and tissues of the body. This pathway is termed the **systemic circulation** *(systemic loop).* These two pathways together with the heart form the basic structure of the cardiovascular system. In this exercise you are challenged to learn some of the major blood vessels that form the circulatory pathways of the human body.

## BEFORE YOU BEGIN

☐ Read the appropriate chapter in your textbook.
☐ Set your learning goals. When you finish this exercise, you should be able to:
  ☐ locate major veins and arteries in models and charts
  ☐ describe the circulatory pathways to and from major body regions
  ☐ describe the fetal circulatory plan and the changes in circulation that occur around the time of birth
☐ Prepare your materials:
  ☐ model of the human body (showing major vessels)
  ☐ charts of human circulatory routes (including fetal circulation)
☐ Read the directions for this exercise *carefully* before starting any procedure.

## *Study* TIPS

Develop a schematic chart of the major arterial branches. Use the arteries that are mentioned in this lab activity. For reference, use your textbook or the *A&P Survival Guide.*

## A. ARTERIES

**Arteries** are blood vessels that conduct blood away from the heart and toward tissues **(Figure 38-1)**. In the pulmonary circulation, **pulmonary arteries** conduct deoxygenated blood to the lungs. In the systemic circulation, the aorta and its branches conduct oxygenated blood toward the systemic tissues. Small arteries are usually called **arterioles.** Arterioles conduct blood into a network of even smaller vessels, or **capillaries.** On a chart, and then in a model of the human body, locate the major human arteries listed. As you do so, note how they form routes to major regions of the body.

☐ 1  Find these portions of the aorta:
  ☐ **Ascending aorta**—First portion, arising from the left ventricle, and the **coronary arteries** branch from this portion of the aorta

☐ **Aortic arch**—Bend of the aorta, just superior to the heart
☐ **Descending aorta**—The remainder of the aorta
☐ **Thoracic aorta**—Portion of the descending aorta that is within the thorax
☐ **Abdominal aorta**—Portion of the descending aorta that is within the abdomen

☐ 2  Locate these arteries of the head and neck:
  ☐ **Brachiocephalic**—First branch from the aortic arch
  ☐ **Right common carotid**—Medial branch of the brachiocephalic artery
  ☐ **Right subclavian**—Lateral branch of the brachiocephalic artery
  ☐ **Left common carotid**—Second branch of the aortic arch
  ☐ **Left subclavian**—Third branch of the aortic arch
  ☐ **Internal carotid**—Branches of the common carotid arteries
  ☐ **External carotid**—Branches of the common carotid arteries
  ☐ **Vertebral**—Branches of the subclavians
  ☐ **Basilar**—Formed by the joining of vertebral arteries
  ☐ **Cerebral arterial circle (of Willis)**—Circular system of arteries around the brain's base, formed by branches of the basilar artery and the internal carotids

☐ 3  Identify these arteries of the upper limb:
  ☐ **Axillary**—Continuation of the subclavian artery inferior to the clavicle
  ☐ **Brachial**—Continuation of the axillary artery in the upper arm
  ☐ **Ulnar**—Medial branch of the brachial artery
  ☐ **Radial**—Lateral branch of the brachial artery

☐ 4  Identify these branches of the descending aorta:
  ☐ **Intercostal**—To the intercostal muscles
  ☐ **Phrenic**—To the diaphragm
  ☐ **Celiac**—To the stomach, pancreas, and liver
  ☐ **Superior mesenteric**—To the small intestine and proximal large intestine
  ☐ **Inferior mesenteric**—To the distal regions of the large intestine
  ☐ **Renal**—To the kidneys
  ☐ **Suprarenal**—To the adrenal glands
  ☐ **Testicular**—To the testis of the male
  ☐ **Ovarian**—To the ovary of the female
  ☐ **Common iliac**—Branches from the inferior end of the descending aorta toward a leg

☐ 5  Identify these arteries of the pelvis and lower limbs:
  ☐ **Internal iliac**—Medial branch of the common iliac artery
  ☐ **External iliac**—Lateral branch of the common iliac artery

Name: _____   Date: _____   Section: _____

## LAB REPORT 38
# The Circulatory Pathway

**Figure 38-1** (from p. 372)

1. _____
2. _____
3. _____
4. _____
5. _____
6. _____
7. _____
8. _____
9. _____
10. _____
11. _____
12. _____
13. _____
14. _____
15. _____
16. _____
17. _____
18. _____
19. _____
20. _____
21. _____
22. _____

**Put in Order** (starting with the left ventricle)

1. left ventricle
2. _____
3. _____
4. _____
5. _____
6. _____
7. _____
8. _____
9. _____
10. _____
11. _____
12. _____
13. _____
14. _____
15. _____
16. _____
17. _____
18. _____

**Identify**

1. _____
2. _____
3. _____
4. _____
5. _____
6. _____
7. _____
8. _____

**Put in Order** (rearrange these structures in the order in which blood passes through them in a circuit from the heart to the foot and back; the first one is done for you)

abdominal aorta
anterior tibial artery
anterior tibial vein
aortic arch
ascending aorta
common iliac vein
common iliac artery
external iliac artery
external iliac vein
femoral artery
femoral vein
foot
inferior vena cava
popliteal artery
popliteal vein
right atrium
thoracic aorta

**Identify** (tell what structure is described in each item)

1. Portion of the aorta in the abdomen
2. Continuation of the subclavian artery inferior to the clavicle
3. Artery that supplies the diaphragm
4. Vein that drains the liver
5. Vessel that conducts blood from digestive organs to the liver
6. Major medial tributary of the external iliac vein
7. Unpaired vein that drains into the posterior aspect of the superior vena cava
8. First major branch of the aortic arch

**Figure 38-2** (from p. 374)

1. _____
2. _____
3. _____
4. _____
5. _____
6. _____
7. _____
8. _____
9. _____
10. _____
11. _____
12. _____
13. _____
14. _____
15. _____
16. _____
17. _____
18. _____
19. _____

**Figure 38-3** (from p. 375)

1. _____
2. _____
3. _____
4. _____
5. _____
6. _____

**Figure 38-5** (from p. 377)

1. _____
2. _____
3. _____
4. _____
5. _____

**Concept Mapping** To learn the concept of how blood flows through various circulatory routes, draw a schematic diagram (of a design of your choice) that clearly shows the structures through which blood flows in the following routes:

1. Hepatic portal circulation
2. Fetal circulation
3. Systemic circulation: to and from any organ of your choice

| VESSEL | TISSUES SUPPLIED OR DRAINED |
|---|---|
| Brachiocephalic artery | |
| Left common carotid artery | |
| Left subclavian artery | |
| Celiac artery | |
| Phrenic arteries | |
| Superior mesenteric artery | |
| Inferior mesenteric artery | |
| Renal artery | |
| Suprarenal artery | |
| External iliac artery | |
| Internal iliac artery | |
| Hepatic vein | |
| External iliac vein | |
| Internal iliac vein | |
| Renal vein | |
| Testicular vein | |
| Ovarian vein | |
| Phrenic vein | |
| Superior mesenteric vein | |
| Inferior mesenteric vein | |
| Gastroepiploic vein | |
| Cystic vein | |

# The Lymphatic System

The **lymphatic system** has several primary functions: it participates in maintaining extracellular fluid balance, it absorbs fats and other substances from the digestive tract, and it functions in the **immune system** defense of the body. To do this, the lymphatic system is composed of a number of vessels and other structures. This exercise presents some of the basics of lymphatic anatomy and physiology.

## BEFORE YOU BEGIN

☐ Read the appropriate chapter in your textbook.
☐ Set your learning goals. When you finish this exercise, you should be able to:
  ☐ describe the major organs of the lymphatic system and find them in a chart
  ☐ identify the features of a lymph node
  ☐ briefly state the function of lymph nodes
☐ Prepare your materials:
  ☐ charts or models of the lymphatic system
  ☐ compound light microscope
  ☐ prepared microslides:
  ☐ lymphatic vessel l.s.
  ☐ lymph node c.s.
  ☐ palatine tonsil c.s.
  ☐ spleen c.s.
  ☐ thymus c.s.
  ☐ colored pencils or pens
☐ Read the directions and safety tips for this exercise *carefully* before starting any procedure.

**Study TIPS**

The best way to understand the role of the lymphatic system is to use the analogy of the system of storm drains and sewage treatment in a city. Check out the complete analogy described in the *A&P Survival Guide*.

## A. OVERVIEW OF THE LYMPHATIC SYSTEM

Use a chart or model of the lymphatic system, along with **Figure 39-1,** to find these gross features:

☐ 1 Locate some of the **lymphatic vessels**. These vessels are similar to veins in the structure of their walls and the presence of many valves. Lymph vessels collect **lymph** from interstitial spaces in the regions served and conduct it toward the blood circulation. Lymph, or *lymphatic fluid*, is an excess of fluid from interstitial areas. The **right lymphatic duct** is a large collecting vessel that receives lymph from the superior right quadrant of the body. The **thoracic duct** similarly conducts lymph received from the rest of the body. Each duct empties into a subclavian vein.

☐ 2 Lymph nodes are small, round organs located at irregular intervals in the network of lymph vessels. Like all *lymphoid* organs, they contain **lymphoid tissue** composed mainly of lymphocytes (a category of white blood cell). They are distributed unevenly, with major aggregations of nodes in the neck (**cervical nodes**), armpit (**axillary nodes**), and groin (**inguinal nodes**).

☐ 3 The **tonsils** are *lymphoid* organs that surround the openings of the mouth, nose, and throat into the lower digestive and respiratory tracts. The **palatine tonsils** are on each lateral side of the opening of the mouth into the throat (see **Figure 39-1, D,** and **Figure 39-2, A**). The **pharyngeal tonsils** are at the top of the throat (*pharynx*) near the posterior of the nasal cavity. The **lingual tonsils** are on the posterior of the tongue's base (see **Figure 39-2, A**).

☐ 4 The **spleen** and **thymus** should already be familiar to you. The spleen is located in the upper left abdominal cavity and the thymus is located in the anterior mediastinum.

> **Hint** A microscopic study of thymic tissue is offered in Lab Exercise 32— Endocrine Glands. •

**Study TIPS**

Think of the lymph node as a filter. Just as an oil filter cleans the oil in a car's engine, lymph nodes clean the debris that accumulates in your lymph.

## B. THE LYMPH NODE

There are numerous lymph nodes throughout the lymphatic system (see **Figure 39-1, A**). Lymph usually passes through one or more lymph nodes on its path toward the blood circulation. The materials in lymph may stimulate lymphocyte development (an immune response) or may be destroyed by *macrophages* in the node. Identify these structures associated with the lymph node in **Figure 39-2, B** and in a chart or model:

☐ 1 **Afferent lymph vessels** conduct lymph into a lymph node, whereas **efferent lymph vessels** conduct lymph out of each node. The terms afferent and efferent apply to several organ systems. One way to distinguish them: afferent–arriving and efferent–exiting.

1
2
3
4
5
6
7
8
9
10

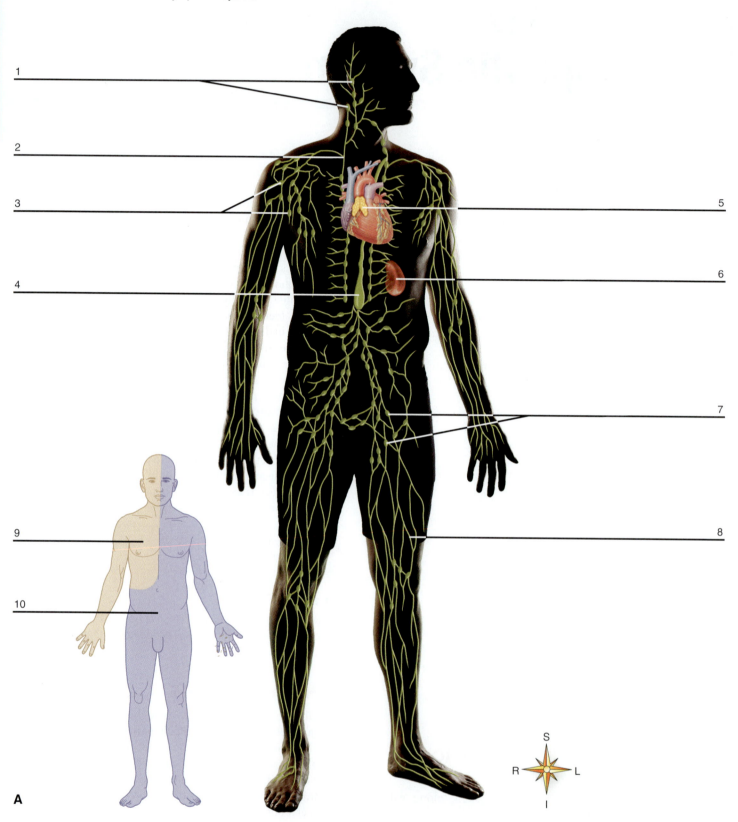

S
R     L
I

A

**FIGURE 39-1**  A, Label the lymphatic structures indicated on the lines provided and on the blanks in the Lab Report at the end of this exercise.

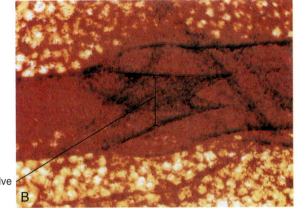

Valve

B

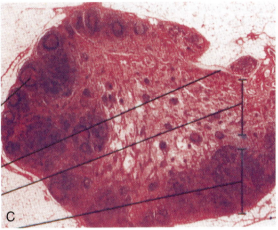

Sinus with
germinal center

Hilum

Medulla

Cortex

C

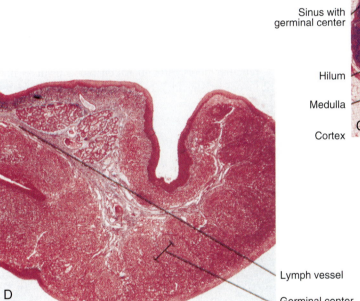

D

Lymph vessel

Germinal center

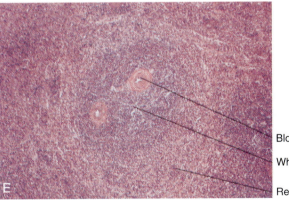

E

Blood vessels

White pulp

Red pulp

**FIGURE 39-1, cont'd** B, Lymphatic vessel. C, Lymph node. D, Palatine tonsil. E, Spleen tissue.

☐ 2  Within the node, lymph tissue often forms dense masses of lymphoid tissue called **lymph nodules.** Each nodule contains a **germinal center** composed of rapidly dividing lymphocytes. Surrounding each node is a space through which lymph circulates; this space is called a **lymph sinus.** The sinuses also contain macrophages.

☐ 3  The fibrous **capsule** functions as the outer boundary and support structure for each lymph node. The inner surface of the capsule has inward extensions that separate the nodules from one another (see **Figure 39-2, B**).

# C. MICROSCOPIC STUDIES OF LYMPHATIC STRUCTURES

To complete our study of the lymphatic system, we will briefly explore some of the microscopic features of structures with which you are already familiar. Please refer to the figures in this exercise and in your textbook as you try to identify the structures described in this section.

> **⓵ safety first**
> Do not forget the rules for safe use of the microscope.  •

☐ 1  **Lymphatic vessel**—As described in Section A, lymphatic vessels are similar to veins in the structure of their walls and the presence of regulatory valves that ensure one-way flow of fluid. Scan your specimen to locate at least one of these valves (see **Figure 39-1, B**).

> **📌 landmark characteristics**
> Compare your specimen with those in **Figure 39-1, B-E**. You will notice relatively thin walls compared with blood vessels of the same size, and more valves than normally seen in veins of the same size.  •

☐ 2  **Lymph node**—See Section A, Step 2, for an overview of the structure of the lymph node. Also see **Figure 39-1, C**, and **Figure 39-2, B**.

☐ 3  **Tonsil**—See Section A, Step 3, for an overview of the structure of tonsils. Also see **Figure 39-1, D**, and **Figure 39-2, A**.

> **📌 landmark characteristics**
> A well-prepared cross-section of a lymph node should show all the features shown in **Figure 39-2, B**. Lymph nodes are often rounded with a distinct capsule that projects inward to separate the tissue of the cortex into sinuses that contain lymph nodules. Each lymph nodule will be seen to contain a pinkish, rounded germinal center, or *follicle*, surrounded by a distinct purple line. Toward the middle of the node, you may see the medulla, which is made up of cords of lymphocytes and plasma cells, with macrophages spanning the spaces between the cords. As **Figure 39-1, D**, shows, a tonsil is essentially a smaller version of a lymph node—that is, a stand-alone nodule with a single germinal center.  •

☐ 4  **Spleen**—The spleen, like the lymph node, has an outer capsule that divides it into subsections **(Figure 39-3)**. Arteries that enter these compartments are surrounded by masses of developing lymphocytes, forming a whitish lymphoid tissue, here called **white pulp.** Surrounding the white pulp is a network of reticular fibers through which blood tissue is flowing. This bloody meshwork is called **red pulp** (see **Figure 39-1, E**).

> **📌 landmark characteristics**
> The splenic capsule and its inward extensions (trabeculae) are normally pale, easy-to-find structures. Within the compartments of the spleen, try to find an area where there are one or more small arteries close together. Surrounding the arteries will be a darkly stained region of cells; this is the white pulp. The red pulp is less densely stained and surrounds the darker white pulp. You may notice some fine reticular fibers in the red pulp.  •

☐ 5  **Thymus**—The thymus has a dual role in that it functions as both a lymphatic organ and an endocrine gland. It serves as a site for the development of T lymphocytes (immune cells) and secretes a hormone that influences lymphocyte development. The gland is surrounded by a *capsule* that extends inward, forming *trabeculae* that divide the thymus into lobules **(Figure 39-3, B)**. Each lobule has a **cortex,** largely filled with lymphocytes, and a medulla with few lymphocytes. In the medulla, small spheres called *thymic (Hassall) corpuscles,* whose function is unclear, can be found (see **Figure 39-4, B**).

# Coloring Exercises: Tonsils and Lymph Node

Use colored pens or pencils to shade in both the figure and the labels. Each red numeral in the figure corresponds to a matching red numeral following the appropriate label.

**Tonsils**
PALATINE TONSILS 1
PHARYNGEAL TONSILS 2
LINGUAL TONSILS 3

**Lymph Node**
AFFERENT VESSEL 1
EFFERENT VESSEL 2

LYMPH NODULE 3
GERMINAL CENTER 4
LYMPH SINUS 5
CAPSULE 6
ARTERY 7
VEIN 8

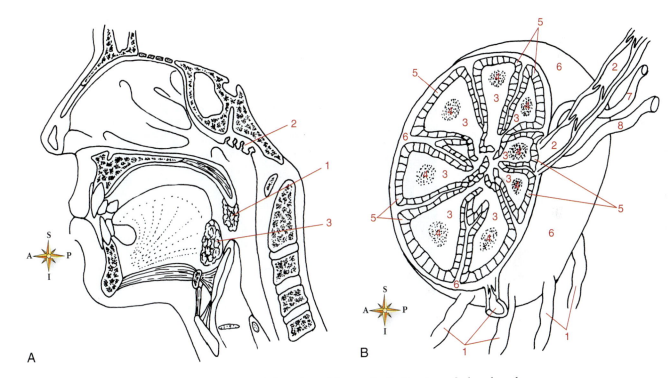

**FIGURE 39-2  A,** Location of the tonsils. **B,** Structure of a lymph node.

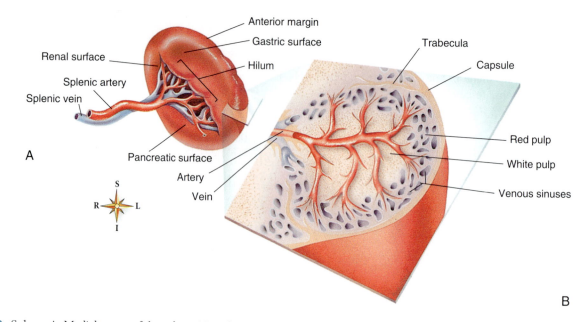

**FIGURE 39-3**   Spleen. **A,** Medial aspect of the spleen. Note the concave surface that fits against the stomach within the abdominopelvic cavity. **B,** Section showing the internal organization of the spleen.

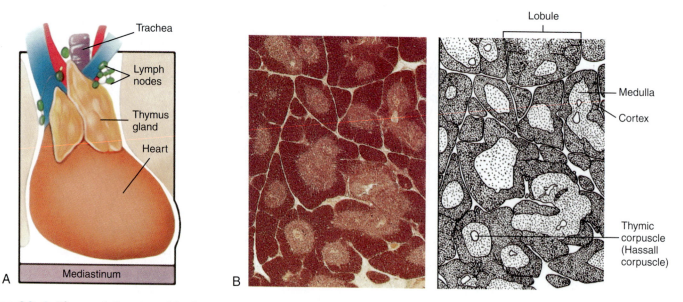

**FIGURE 39-4**   Thymus. **A,** Location of the thymus within the mediastinum. **B,** Microscopic structure of the thymus (micrograph and drawing) showing several lobules, each with a cortex and a medulla.

### 📌 *landmark characteristics*

The thymic capsule and its inward extensions (trabeculae) are normally pale, easy-to-find structures. Near these walls, the cortex can be seen as a region of darkly stained lymphocytes. Farther away from the walls, the more lightly stained medullary region can be seen. Occasional thymic corpuscles (circles with a nonuniform interior) can sometimes be seen.   •

# *clinical application: immune system and autoimmunity disease*

Lymphocytes may be involved in two major types of responses to the presence of potentially threatening foreign substances. In **antibody-mediated immunity,** *B lymphocytes* may produce antibodies that react with specific antigens on an invading cell or molecule. In **cell-mediated immunity,** *T lymphocytes* secrete **lymphokines** that signal other immune responses and often destroy antigen-containing cells directly.

The antibody-antigen reactions associated with immune responses have been used by clinical biologists for years. Several examples are given. Think about each example and answer the questions.

1. Biologists often use antibodies to test for the presence of certain antigens in a particular substance. In Lab Exercise 33—Hormones and Lab Exercise 34—Blood, you witnessed demonstrations of this technique. For each antigen listed, indicate the substance tested (e.g., blood, urine) and give a brief summary of the antibody-antigen reaction involved.

hCG ANTIGEN:

A and B ANTIGENS:

D ANTIGEN:

2. *Rheumatoid arthritis* is an inflammatory disease affecting joint tissues. This disease is known to have an *autoimmune* component. Autoimmunity is an immune response inappropriately directed toward normal self-antigens. Many persons with rheumatoid arthritis have *rheumatoid factor (RF)* present in their blood. RF is an abnormal antibody. Not all persons with the disease have RF in their plasma, and RF is known to be present in conditions other than rheumatoid arthritis. Based on this information and your previous study, outline a simple test to detect the presence of RF in plasma. If your test is used to screen for rheumatoid arthritis, does a positive result (i.e., RF is present) mean that the person definitely has the disease? Why or why not?

Name: _____ Date: _____ Section: _____

**LAB REPORT 39**
# The Lymphatic System

**Figure 39-1** (from p. 382)

1. _____
2. _____
3. _____
4. _____
5. _____
6. _____
7. _____
8. _____

**Ducts** (from p. 382)

9. _____
10. _____

**Multiple Choice**

1. _____
2. _____
3. _____
4. _____
5. _____
6. _____
7. _____

**Multiple Choice** (choose the best response—only one is correct)

1. Lymphatic tissue contains
   a. mainly lymphocytes
   b. mainly erythrocytes
   c. mainly monocytes
2. The large lymphatic organ found in the left hypochondriac region is the
   a. thymus
   b. palatine tonsil
   c. spleen
   d. lingual tonsil
   e. inguinal lymph node
3. The lymph from tissues of the lower limbs is conducted by the
   a. right lymphatic duct
   b. thoracic duct
4. Lymph nodes function to
   a. provide a site for lymphocyte reproduction
   b. anchor lymphatic vessels
   c. provide a site for phagocytosis of foreign cells or debris
   d. a and b are correct
   e. a and c are correct
5. Lymph vessels that conduct lymph into a lymph node are termed
   a. efferent lymphatic vessels
   b. efferent arterioles
   c. afferent arterioles
   d. afferent lymphatic vessels
   e. none of the above
6. The lymphatic system functions to
   a. maintain extracellular fluid balance
   b. collect excess fluid from the interstitial spaces
   c. defend the body against foreign cells and molecules
   d. a and c are correct
   e. a, b, and c are correct
7. Lymph drains directly into the blood at
   a. the right subclavian vein
   b. the left subclavian vein
   c. the subclavian arteries
   d. a and b are correct
   e. none of the above

| LYMPHATIC STRUCTURE | LOCATION | FUNCTION(S) |
|---|---|---|
| Lymphatic capillaries | | |
| Lymphatic vessels | | |
| Right lymphatic duct | | |
| Thoracic duct | | |
| Lymph node | | |
| Palatine tonsils | | |
| Pharyngeal tonsils | | |
| Lingual tonsils | | |
| Spleen | | |
| Thymus | | |

## Sketch (draw a cross-section of a lymph node and label its parts)

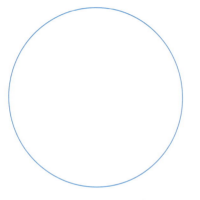

Specimen: *lymphatic vessel l.s.*
Total Magnification: _____

Specimen: *lymph node c.s.*
Total Magnification: _____

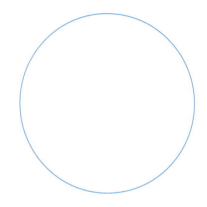

Specimen: *palatine tonsil c.s.*
Total Magnification: _____

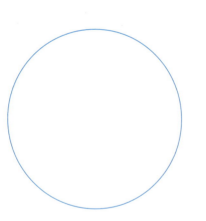

Specimen: *spleen c.s*
Total Magnification: _____

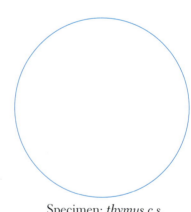

Specimen: *thymus c.s.*
Total Magnification: _____

# LAB EXERCISE 40
# Dissection: Cardiovascular and Lymphatic Systems

In this exercise, you continue your study of cardiovascular and lymphatic anatomy by dissecting a whole preserved specimen.

Activity A provides directions for studying the cardiovascular and lymphatic anatomy of the cat. Activity B is an alternate activity, providing directions for studying the cardiovascular and lymphatic anatomy of the fetal pig.

## BEFORE YOU BEGIN

☐ Read the appropriate chapter in your textbook.
☐ Set your learning goals. When you finish this exercise, you should be able to:
  ☐ dissect the cardiovascular and lymphatic anatomy of a preserved cat or fetal pig
  ☐ identify the following in a dissected mammalian specimen:
    ☐ heart and its major features
    ☐ major arteries
    ☐ major veins
    ☐ lymph nodes
☐ Prepare your materials:
  ☐ preserved (injected) cat or fetal pig
  ☐ dissection tools and trays
  ☐ storage container (if specimen is to be reused)
  ☐ protective gear: gloves, safety eyewear, and apron
☐ Read the directions and safety tips for this exercise *carefully* before starting any procedure.

### *Study* TIPS

As in previous dissections, it would be helpful to use references of the human circulatory and lymphatic systems. You may want to use your cell phone to photograph your specimens. Afterward, you can use this photo as a resource for the circulatory and lymphatic systems.  •

## A. CARDIOVASCULAR AND LYMPHATIC SYSTEMS OF THE CAT

This activity challenges you to identify the major systemic blood vessels of the cat, features of the cat's heart, and lymphoid organs inside the cat's body. To find these structures, you must cut open the cat's thoracic and abdominopelvic cavities. Assuming you have already removed the cat's skin (Lab Exercise 10—The Skin), you can simply cut through the ventral aspect of the thoracic and abdominal musculature. Do this by inserting your scissors into the ventral abdominal wall (without damaging any internal organs), lifting up the scissors, and cutting along the midline toward the head and tail. You will need to cut through the costal cartilage that connects the ribs to the sternum. Transverse cuts from this initial cut will produce "doors" to the internal body cavities that can be pulled back to expose the viscera. You may have to cut the ribs to see the thoracic viscera. Carefully move the viscera out of the way (without cutting or removing them) to see the structures indicated in this activity.

### ⚠ safety first

Observe the usual precautions when working with a preserved specimen. Heed the safety advice accompanying preservatives used with your specimen. Use protective gloves while handling your specimen. Avoid injury with dissection tools. Use safety goggles to avoid injury during dissections. Dispose of or store your specimen as instructed.  •

### Hint

In double-injected specimens, *arteries* are filled with red latex and *veins* are filled with blue latex. This makes these vessels easy to identify. In specimens that are not injected, you can often distinguish veins and arteries by the thickness of their walls—veins often have thinner walls than arteries. Veins and arteries can also be distinguished by their *locations*—systemic veins lead toward a vena cava, and systemic arteries branch from the aorta.  •

☐ 1 Identify the **aorta** extending from the **heart**. Distinguish these regions of the aorta:
  ☐ Ascending aorta
  ☐ Aortic arch
  ☐ Descending aorta
    ☐ Thoracic aorta
    ☐ Abdominal aorta

> **Hint** Figures 40-1 through 40-4 illustrate many of the structures of the cat listed in this activity. •

☐ 2 Identify these branches of the ascending aorta (see **Figure 40-1**):
  ☐ Coronary arteries

☐ 3 Identify these branches of the aortic arch:
  ☐ Brachiocephalic artery (on right side only)
  ☐ Subclavian arteries
  ☐ Axillary arteries
  ☐ Brachial arteries
  ☐ Vertebral arteries
  ☐ Costocervical arteries
  ☐ Common carotid arteries

☐ 4 Identify these branches of the thoracic aorta:
  ☐ Intercostal arteries
  ☐ Anterior phrenic arteries

☐ 5 Identify these branches of the abdominal aorta (see **Figure 40-2**):
  ☐ Posterior phrenic arteries
  ☐ Celiac artery
  ☐ Splenic artery
  ☐ Anterior mesenteric artery
  ☐ Suprarenal arteries
  ☐ Renal arteries
  ☐ Gonadal (testicular or ovarian) arteries
  ☐ Lumbar arteries

☐ Posterior mesenteric artery
☐ External iliac arteries
☐ Internal iliac arteries
☐ Femoral arteries

☐ 6 Identify the large **anterior vena cava**, which corresponds to the *superior vena cava* in the human. Identify these tributaries of the anterior vena cava in the cat (see **Figure 40-3**):
  ☐ Azygous vein
  ☐ Internal mammary veins
  ☐ Brachiocephalic veins
  ☐ Axillary veins
  ☐ Subclavian veins
  ☐ Subscapular veins
  ☐ Transverse scapular veins
  ☐ Brachial veins
  ☐ Long thoracic veins
  ☐ Internal jugular veins
  ☐ External jugular veins
  ☐ Transverse jugular vein
  ☐ Posterior facial veins
  ☐ Anterior facial veins
  ☐ Submental veins

☐ 7 Identify the large **posterior vena cava**, which corresponds to the *inferior vena cava* in the human. Identify these tributaries of the posterior vena cava in the cat (see **Figure 40-4**):
  ☐ Adrenolumbar veins
  ☐ Renal veins
  ☐ Common iliac veins
  ☐ External iliac veins
  ☐ Internal iliac veins
  ☐ Femoral veins
  ☐ Greater saphenous veins

☐ 8 Examine the heart of the cat. Identify as many features as you can, using the information in Lab Exercise 35—Structure of the Heart as a guide.

☐ 9 Locate the **spleen**, a lymphoid organ near the stomach. Try to find one or more **lymph nodes** or **lymphatic vessels**.

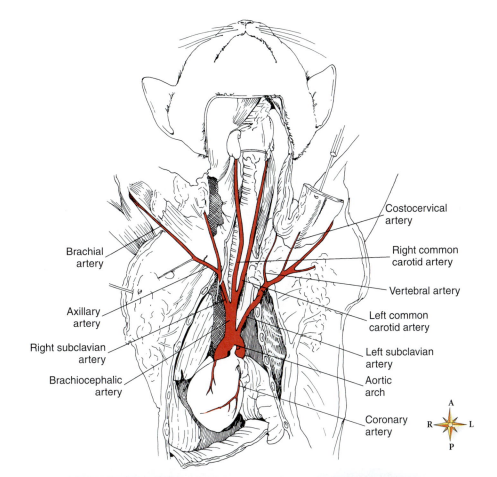

Costocervical artery

Brachial artery

Right common carotid artery

Vertebral artery

Axillary artery

Left common carotid artery

Right subclavian artery

Left subclavian artery

Brachiocephalic artery

Aortic arch

Coronary artery

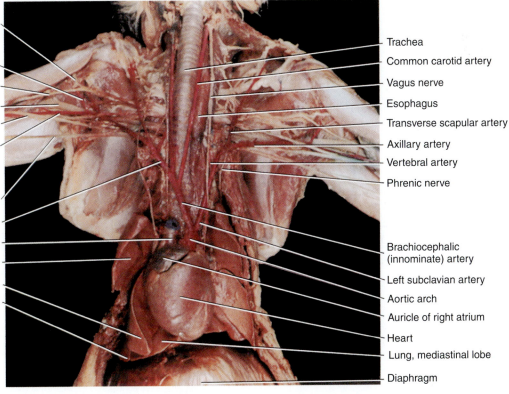

Thoraco-acromial nerve

Thoraco-acromial artery

Musculocutaneous nerve

Radial nerve

Median nerve

Brachial artery

Ulnar nerve

Right subclavian artery

Anterior vena cava (cut)

Lung, anterior lobe

Lung, middle lobe

Lung, posterior lobe

Trachea

Common carotid artery

Vagus nerve

Esophagus

Transverse scapular artery

Axillary artery

Vertebral artery

Phrenic nerve

Brachiocephalic (innominate) artery

Left subclavian artery

Aortic arch

Auricle of right atrium

Heart

Lung, mediastinal lobe

Diaphragm

**FIGURE 40-1** Arteries of the cat. Thorax, ventral view.

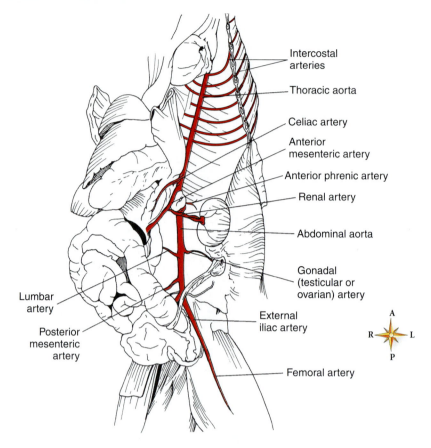

Intercostal arteries
Thoracic aorta
Celiac artery
Anterior mesenteric artery
Anterior phrenic artery
Renal artery
Abdominal aorta
Gonadal (testicular or ovarian) artery
External iliac artery
Femoral artery
Lumbar artery
Posterior mesenteric artery

**FIGURE 40-2** Arteries of the cat. Thorax and abdomen, ventral view.

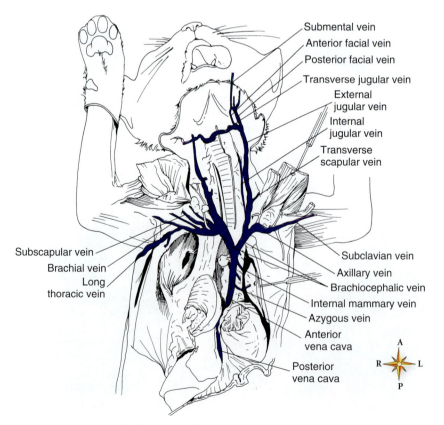

Submental vein
Anterior facial vein
Posterior facial vein
Transverse jugular vein
External jugular vein
Internal jugular vein
Transverse scapular vein
Subscapular vein
Brachial vein
Long thoracic vein
Subclavian vein
Axillary vein
Brachiocephalic vein
Internal mammary vein
Azygous vein
Anterior vena cava
Posterior vena cava

**FIGURE 40-3** Veins of the cat. Neck and thorax, ventral view.

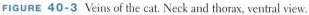

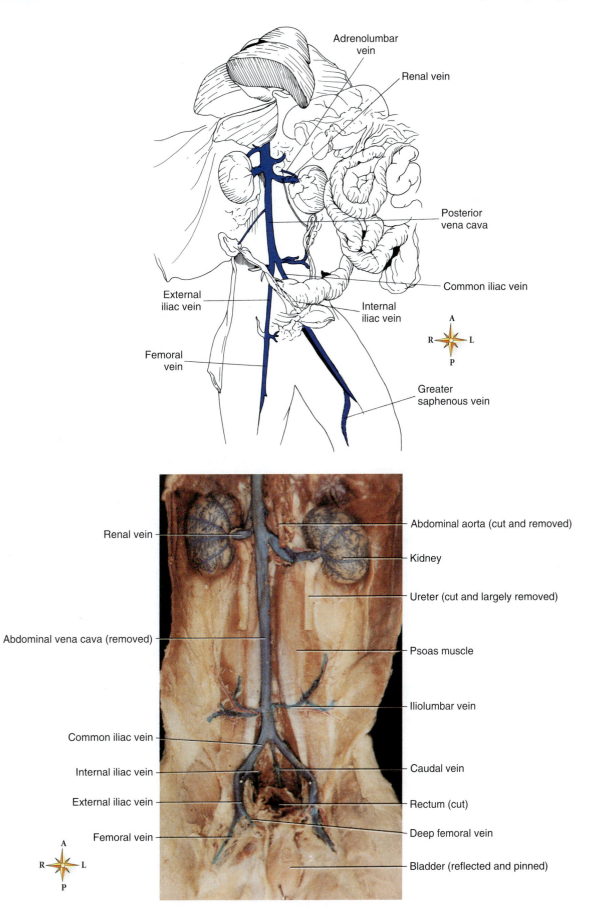

**FIGURE 40-4** Veins of the cat. Abdomen, ventral view.

# B. CARDIOVASCULAR AND LYMPHATIC SYSTEMS OF THE FETAL PIG

This activity challenges you to identify the major systemic blood vessels of the fetal pig, features of the heart, and lymphoid organs inside the fetal pig's body. To find these structures, you must cut open your specimen's thoracic and abdominopelvic cavities. Assuming you have already removed the skin (Lab Exercise 10—The Skin), you can simply cut through the ventral aspect of the thoracic and abdominal musculature. Do this by inserting your scissors into the ventral abdominal wall (without damaging any internal organs), lifting up the scissors, and cutting along the midline toward the head and tail. You will need to cut through the costal cartilage that connects the ribs to the sternum. Transverse cuts from this initial cut will produce "doors" to the internal body cavities that can be pulled back to expose the viscera. You may have to cut the ribs to see the thoracic viscera. Carefully move the viscera out of the way (without cutting or removing them) to see the structures indicated in this activity.

**Hint ▶** In double-injected specimens, *arteries* are filled with red latex and *veins* are filled with blue latex. This makes these vessels easy to identify. In specimens that are not injected, you can often distinguish veins and arteries by the thickness of their walls—veins often have thinner walls than arteries. Veins and arteries can also be distinguished by their *locations*—systemic veins lead toward a vena cava, and systemic arteries branch from the aorta.  •

☐ 1 Identify the **aorta** extending from the **heart**. Distinguish these regions of the aorta:
- ☐ **Ascending aorta**
- ☐ **Aortic arch**
- ☐ **Descending aorta**
  - ☐ **Thoracic aorta**
  - ☐ **Abdominal aorta**

**Hint ▶** Figures 40-5 through 40-7 illustrate many of the structures of the fetal pig listed in this activity.  •

☐ 2 Identify these branches of the ascending aorta (see **Figure 40-5**):
- ☐ **Coronary arteries**

☐ 3 Identify these branches of the aortic arch:
- ☐ **Brachiocephalic artery** (on right side only)
- ☐ **Subclavian arteries**
- ☐ **Axillary arteries**
- ☐ **Brachial arteries**
- ☐ **Common carotid arteries**

☐ 4 Identify these branches of the thoracic aorta:
- ☐ **Intercostal arteries**

☐ 5 Identify these branches of the abdominal aorta (see **Figure 40-6**):
- ☐ **Celiac artery**
- ☐ **Splenic artery**
- ☐ **Anterior mesenteric artery**
- ☐ **Renal arteries**
- ☐ **Gonadal (testicular or ovarian) arteries**
- ☐ **Posterior mesenteric artery**
- ☐ **Umbilical arteries**
- ☐ **External iliac arteries**
- ☐ **Internal iliac arteries**
- ☐ **Femoral arteries**

☐ 6 Identify the large **anterior vena cava**, which corresponds to the *superior vena cava* in the human. Identify these tributaries of the anterior vena cava in the fetal pig (see **Figure 40-7**):
- ☐ **Internal mammary veins**
- ☐ **Brachiocephalic veins**
- ☐ **Axillary veins**
- ☐ **Subclavian veins**
- ☐ **Internal jugular veins**
- ☐ **External jugular veins**

☐ 7 Identify the large **posterior vena cava**, which corresponds to the *inferior vena cava* in the human. Identify these tributaries of the posterior vena cava in the fetal pig (see **Figure 40-6**):
- ☐ **Renal veins**
- ☐ **External iliac veins**
- ☐ **Internal iliac veins**
- ☐ **Femoral veins**

☐ 8 Examine the heart of the fetal pig. Identify as many features as you can, using the information in Exercise 35—Structure of the Heart as a guide.

☐ 9 Locate the **spleen**, a lymphoid organ near the stomach. Try to find one or more **lymph nodes** or **lymphatic vessels**.

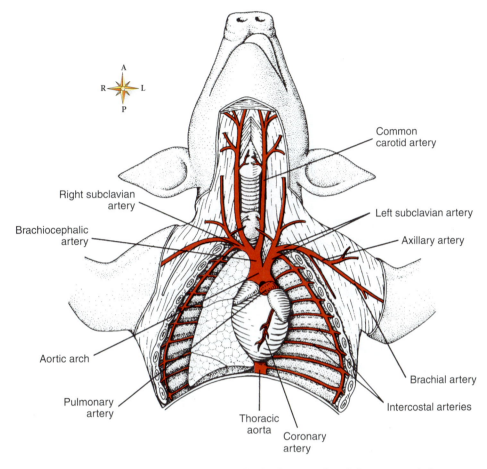

**FIGURE 40-5** Arteries of the fetal pig. Neck and thorax, ventral view.

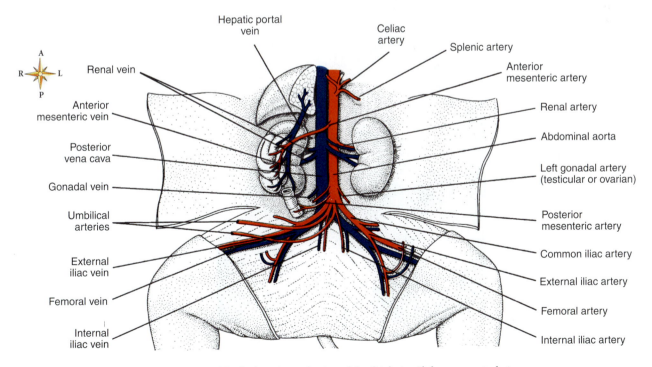

**FIGURE 40-6** Arteries and veins of the fetal pig. Abdomen, ventral view.

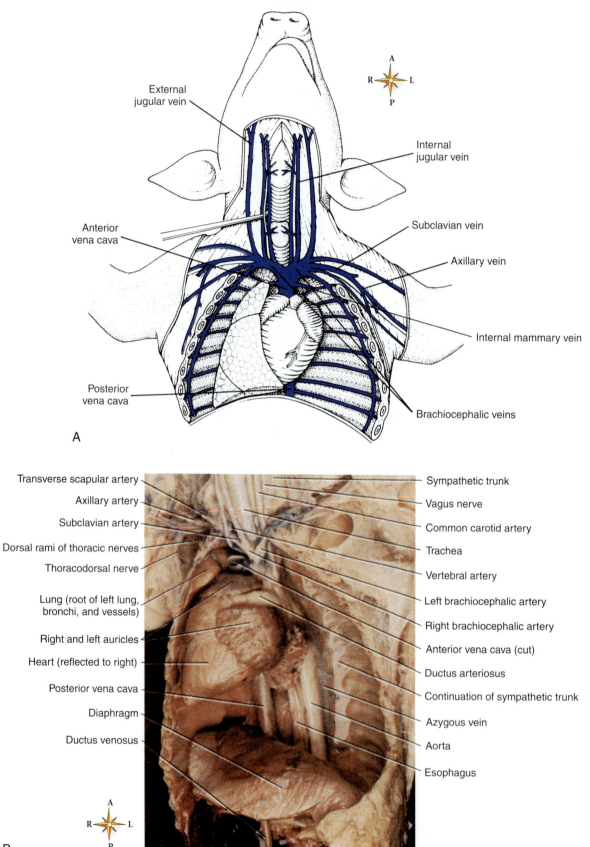

**FIGURE 40-7** **A,** Veins of the fetal pig. Neck and thorax, ventral view. **B,** Vessels of the fetal pig. Neck and thorax, ventral view.

Name: _____   Date: _____   Section: _____

**LAB REPORT 40**

# Dissection: Cardiovascular and Lymphatic Systems

## A. Cat Dissection Checklist

☐ Arteries

☐ aorta

☐ ascending aorta

☐ coronary arteries

☐ aortic arch

☐ brachiocephalic artery (on right side only)

☐ subclavian arteries

☐ axillary arteries

☐ brachial arteries

☐ vertebral arteries

☐ costocervical arteries

☐ common carotid arteries

☐ thoracic aorta

☐ intercostal arteries

☐ anterior phrenic arteries

☐ abdominal aorta

☐ posterior phrenic arteries

☐ celiac artery

☐ splenic artery

☐ anterior mesenteric artery

☐ suprarenal arteries

☐ renal arteries

☐ gonadal (testicular or ovarian) arteries

☐ lumbar arteries

☐ posterior mesenteric artery

☐ external iliac arteries

☐ internal iliac arteries

☐ femoral arteries

☐ Veins

☐ anterior vena cava

☐ azygous vein

☐ internal mammary veins

☐ brachiocephalic veins

☐ axillary veins

☐ subclavian veins

☐ subscapular veins

☐ transverse scapular veins

☐ brachial veins

☐ long thoracic veins

☐ internal jugular veins

☐ external jugular veins

☐ transverse jugular vein

☐ posterior facial veins

☐ anterior facial veins

☐ submental veins

☐ posterior vena cava

☐ adrenolumbar veins

☐ renal veins

☐ common iliac veins

☐ external iliac veins

☐ internal iliac veins

☐ femoral veins

☐ greater saphenous veins

☐ Other organs

☐ heart

☐ spleen

☐ lymph nodes

☐ lymphatic vessels

☐ **2** Identify these features of the **pharynx**, or throat **(Figure 41-2)**:
  ☐ **Nasopharynx**—The upper of three regions of the pharynx, this one is posterior to the nasal cavity. The **uvula** is an extension of the floor of the nasopharynx, or **soft palate**, and serves as the inferior boundary landmark of the nasopharynx. *The pharyngeal tonsils* are on the posterior wall, and the lateral walls have the inferior opening of the *auditory (eustachian) tube* from the middle ear.
  ☐ **Oropharynx**—The middle portion of the pharynx, it is posterior to the oral cavity. The *palatine* and *lingual tonsils* are found here.

☐ **Laryngopharynx**—The most inferior portion of the pharynx, it is posterior to the larynx.
☐ **3** Identify these features of the voice box, or **larynx**:
  ☐ **Laryngeal cartilages**—Nine pieces of cartilage (6 in pairs and 3 single pieces) that fit together to form the wall of the larynx: **thyroid, cricoid, epiglottis, cuneiforms** (pair), **corniculates** (pair), and **arytenoids** (pair)
  ☐ **Vestibular folds** (false vocal cords)—Superior of two pairs of ligaments stretched across the lateral portions of the larynx cavity
  ☐ **Vocal folds** (true vocal cords)—Inferior of two pairs of laryngeal ligaments

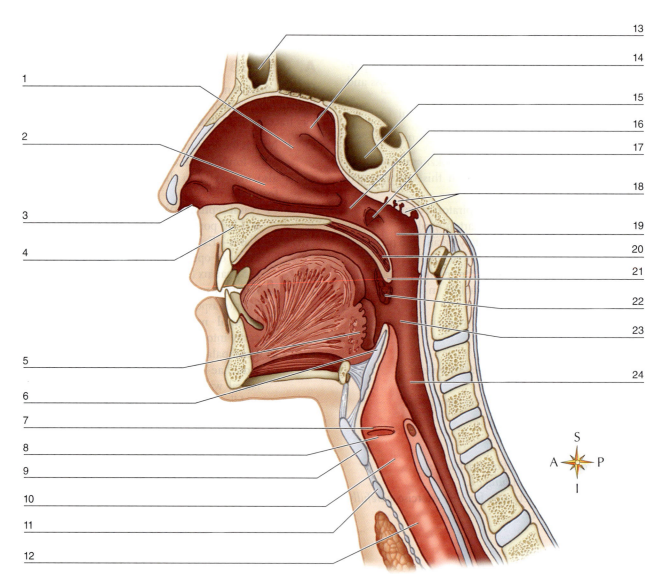

**FIGURE 41-1** Midsagittal section of the head, showing respiratory structures. Label the structures indicated on the lines provided and on the blanks in the Lab Report at the end of this exercise.

# Coloring Exercises: Respiratory Anatomy

Use colored pens or pencils to shade in both the figure and the labels. Each red numeral in the figure corresponds to a matching red numeral following the appropriate label.

NASAL CAVITY 1
NASOPHARYNX 2
OROPHARYNX 3
LARYNGOPHARYNX 4
LARYNX 5
TRACHEA 6
PRIMARY BRONCHI 7
SECONDARY BRONCHI 8
TERTIARY BRONCHI 9
BRONCHIOLE 10

TERMINAL BRONCHIOLE 11
RESPIRATROY BRONCHIOLE 12
ALVEOLAR DUCT 13
ALVEOLUS 14

**Lobes (Shade Lightly)**

RIGHT SUPERIOR LOBE 15
RIGHT MIDDLE LOBE 16
RIGHT INFERIOR LOBE 17
LEFT SUPERIOR LOBE 18
LEFT INTERIOR LOBE 19

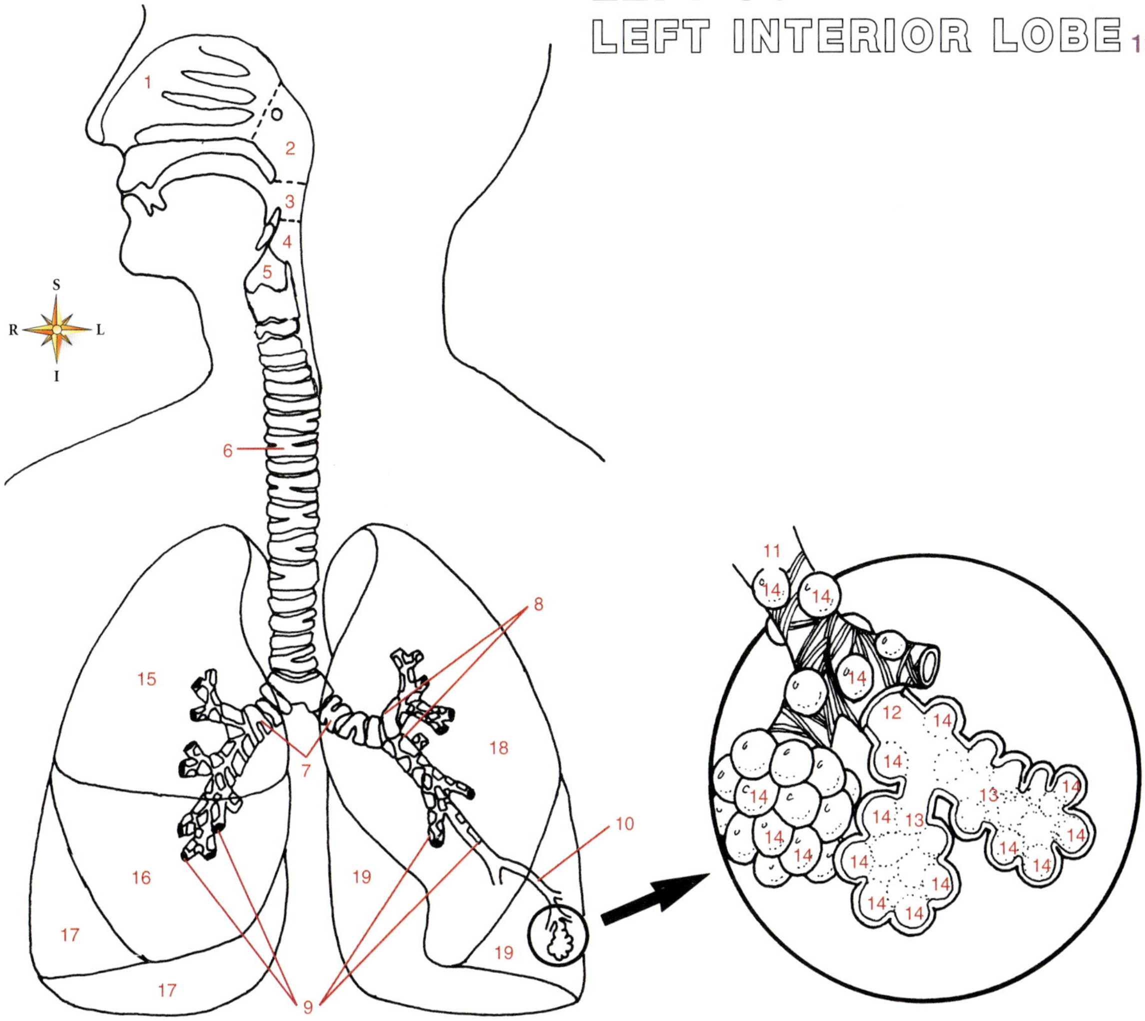

**FIGURE 41-2** Structural plan of the respiratory system.

# Figure 41-1 (from p. 402)

1. _____
2. _____
3. _____
4. _____
5. _____
6. _____
7. _____
8. _____
9. _____
10. _____
11. _____
12. _____
13. _____
14. _____
15. _____
16. _____
17. _____
18. _____
19. _____
20. _____
21. _____
22. _____
23. _____
24. _____

## Put in Order

1. _____
2. _____
3. _____
4. _____
5. _____
6. _____
7. _____
8. _____
9. _____

## Put in Order (rearrange these structures in the order through which air passes during inspiration)

larynx
nasal cavity
pharynx
primary bronchi
respiratory bronchioles
secondary bronchi
terminal bronchioles
tertiary bronchi
trachea

# LAB EXERCISE 42
# Dissection: Respiratory System

In this exercise, you continue your study of respiratory anatomy by dissecting a whole preserved specimen.

Activity A provides directions for studying the respiratory anatomy of the cat. Activity B is an alternate activity, providing directions for studying the respiratory anatomy of the fetal pig. Activity C is also an alternate activity that asks you to dissect a sheep pluck, or lower respiratory tract and heart of a sheep.

## BEFORE YOU BEGIN

☐ Read the appropriate chapter in your textbook.
☐ Set your learning goals. When you finish this exercise, you should be able to:
  ☐ dissect the respiratory anatomy of a preserved cat, fetal pig, or sheep
  ☐ identify the following in a dissected mammalian specimen:
    ☐ larynx
    ☐ trachea and bronchial tree
    ☐ lungs
    ☐ diaphragm
☐ Prepare your materials:
  ☐ preserved cat, fetal pig, or sheep pluck
  ☐ dissection tools and trays
  ☐ protective gear: gloves, safety eyewear, and apron
  ☐ storage container (if specimen is to be reused)
  ☐ dissection microscope (optional)
☐ Read the directions and safety tips for this exercise *carefully* before starting any procedure.

> ⚠ **safety first**
> Observe the usual precautions when working with a preserved specimen. Heed the safety advice accompanying preservatives used with your specimen. Use protective gloves while handling your specimen. Avoid injury with dissection tools. Use safety goggles to avoid injury during dissections. Dispose of or store your specimen as instructed. •

## *Study* TIPS

As you progress through this dissection, compare your specimen with a diagram or illustration of the human respiratory tract. For reference, use your textbook or the *A&P Survival Guide*. You may want to use your cell phone to take a picture of your specimen and use it later to identify the respiratory structures.

## A. RESPIRATORY ANATOMY OF THE CAT

This activity challenges you to identify the major respiratory structures of the cat. Assuming you have already removed the cat's skin (Lab Exercise 10—The Skin) and opened the thoracic cavity (Lab Exercise 40—Dissection: Cardiovascular and Lymphatic Systems),

you simply have to cut through the muscles of the neck and head to expose internal features of the anterior airway (equivalent to the upper respiratory tract in humans). Be careful as you do this to avoid damaging the specimen to the point that you really cannot tell where the structures are normally located in relation to one another. Remember to carefully move the viscera out of the way (without cutting or removing them) to see the structures indicated in this activity.

☐ 1 Identify the **external nares,** the openings of the **nasal cavity.** Notice that a **nasal septum** divides the nasal cavity into two air passages. If your instructor directs you to do so, cut into the nose and mouth of the cat to expose the nasal cavity and **pharynx.** In the nasal cavity, try to identify the nasal **turbinates** *(conchae)* projecting from the lateral walls. Can you identify the three major regions of the pharynx: *nasopharynx, oropharynx,* and *laryngopharynx*?

☐ 2 Locate and identify the **larynx,** or voice box, deep in the neck. The **epiglottis, thyroid cartilage,** and **cricoid cartilage** of the larynx are easily visible on the ventral aspect.

☐ 3 Identify the **trachea** extending posteriorly from the larynx. Notice the cartilage rings that support this structure.

> **Hint** ▶ **Figure 42-1** illustrates many of the structures of the cat listed in this activity. •

☐ 4 Identify the right and left **primary bronchi.** Notice that each extends to a different **lung,** where it splits to form several **secondary bronchi.**

☐ 5 Each secondary bronchus serves a distinct **lobe** of a lung. Identify the following lobes in each lung of your specimen:
  ☐ **Anterior lobe** (left and right)
  ☐ **Middle lobe** (left and right)
  ☐ **Posterior lobe** (left and right)
  ☐ **Mediastinal lobe** (right only)

☐ 6 If your instructor directs you to do so, slice into the tissue of one of the lobes and remove a small section of tissue. Examine the internal structure of this tissue with a dissection microscope or hand lens to reveal air passages that constitute distal elements of the **bronchial tree** and gas-exchange tissues. You may make a deeper cut into the lobe to reveal other elements of the bronchial airways.

☐ 7 Locate the following structures of the wall of the thoracic cavity:
  ☐ **Intercostal muscles**
  ☐ **Thoracic inlet**
  ☐ **Parietal pleura**
  ☐ **Visceral pleura**
  ☐ **Pleural space**
  ☐ **Diaphragm**
  ☐ **Phrenic nerve (innervates diaphragm)**

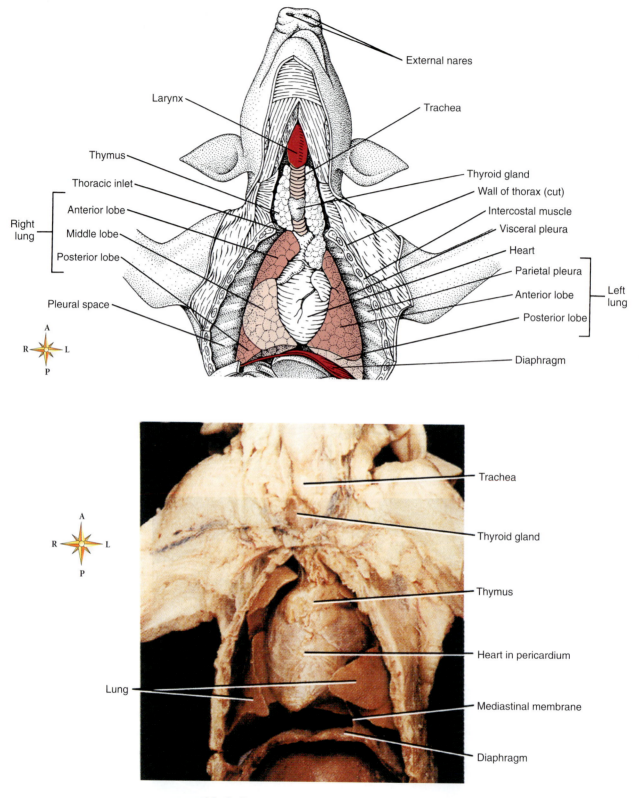

**FIGURE 42-2** Respiratory anatomy of the fetal pig. Ventral view.

## C. RESPIRATORY ANATOMY OF THE SHEEP

This activity challenges you to identify the major respiratory structures in a *sheep pluck* specimen. This specimen includes a portion of the respiratory system, the heart, and other nearby structures removed from the carcass of a sheep. The organs of the sheep pluck are similar in size to the adult human (depending on the particular sheep and the particular human, of course).

☐ 1 Locate and identify the **larynx**, or voice box, at the anterior end of the sheep pluck.

☐ 2 Identify the **trachea** extending posteriorly from the larynx. Notice the cartilage rings that support this structure.

> **Hint**  **Figure 42-3** illustrates many of the structures of the sheep pluck listed in this activity.  •

☐ 3 Identify the right and left **primary bronchi**. Notice that each extends to a different **lung**, where it splits to form several **secondary bronchi**.

☐ 4 Each secondary bronchus serves a distinct **lobe** of a lung. Identify the lobes in each lung of your specimen. How many are there in each lung? How does this compare to the human?

☐ 5 If your instructor directs you to do so, slice into the tissue of one of the lobes and remove a small section of tissue. Examine the internal structure of this tissue with a dissection microscope or hand lens to reveal air passages that constitute distal elements of the **bronchial tree** and gas-exchange tissues. You may make a deeper cut into the lobe to reveal other elements of the bronchial airways.

After exercise (0 min):

| VOLUME OR CAPACITY | EXPECTED VALUE[a] | ACTUAL VALUE | DIFFERENCE | INTERPRETATION |
|---|---|---|---|---|
| TV | | | | |
| ERV | | | | |
| VC | | | | |
| IRV | | | | |

[a]Use the results from the first table on the previous page as your expected results. Be sure to use the correct units of volume when reporting your values.

| VOLUME OR CAPACITY | EXPECTED VALUE* | ACTUAL VALUE | DIFFERENCE | INTERPRETATION |
|---|---|---|---|---|
| TV | | | | |
| ERV | | | | |
| VC | | | | |
| IRV | | | | |

[a]Use the results from the first table on the previous page as your expected results. Be sure to use the correct units of volume when reporting your values.

Discuss your results (comparing resting values with values after exercise):

# Digestive Structures

The **digestive tract** is a series of hollow organs through which food passes: **mouth, pharynx, esophagus, stomach, small intestine,** and **large intestine.** Each portion is specialized for one or more aspects of the three major functions of the **digestive system:** *secretion, digestion,* and *absorption.* Accessory organs such as the **salivary glands, liver, gallbladder,** and **pancreas** have ducts that lead into the digestive tract and thus support digestive function.

This exercise challenges you to explore the structure and some of the essential functions of the digestive system.

## BEFORE YOU BEGIN

☐ Read the appropriate chapter in your textbook.
☐ Set your learning goals. When you finish this exercise, you should be able to:
  ☐ describe the structure of digestive organs and locate them in models and charts
  ☐ identify the principal function of each major digestive organ
  ☐ describe the basic histology of the gastrointestinal (GI) wall
☐ Prepare your materials:
  ☐ model of the human torso (dissectible) or chart
  ☐ human skull (with teeth intact)
  ☐ model or specimen of a tooth section
  ☐ compound light microscope
  ☐ prepared microslides:
    ☐ stomach wall c.s.
    ☐ small intestine wall c.s.
    ☐ salivary gland c.s.
  ☐ colored pencils or pens
☐ Read the directions and safety tips for this exercise *carefully* before starting any procedure.

## Study TIPS

Develop a schematic diagram that traces the pathway of food through the entire digestive tract. Include in your design what happens to food as it passes each digestive structure.

## A. WALL OF THE DIGESTIVE TRACT

The wall of the digestive tract **(Figure 44-1)** forms a hollow tube continuous with the external environment. Food passes along the tube in a process usually termed **motility.** The wall of the tract is composed of four layers (tunics). They are:

☐ **Mucosa**—Inner lining, consisting of mucous epithelium, connective tissue *(lamina propria),* and a thin layer of smooth muscle *(muscularis mucosae).*
☐ **Submucosa**—Thick layer of connective tissue just superficial to the mucosa; the submucosa includes blood vessels, a network

of nerve fibers called the *submucosal plexus,* and *submucosal glands.*
☐ **Muscularis**—Layers of smooth muscle surrounding the submucosa—usually an inner *circular layer* and an outer *longitudinal layer* (the stomach also has a middle *oblique layer*); the muscularis has a network of nerve fibers called the *myenteric plexus.*
☐ **Serosa (adventitia)**—Most superficial layer of the digestive wall, composed entirely of connective tissue (adventitia) or of connective tissue fused to serous epithelial tissue (serosa); the serous membrane extends away from the GI wall in folds called *mesenteries;* digestive glands with ducts that lead through the GI wall and into the *lumen* (hollow part) of the GI tract may lie outside the serosa or adventitia.

☐ 1  Locate the digestive tract in a dissectible model of the human torso. Some portions may have the wall cut in a cross-section. If so, try to identify each of the four layers previously described. How does each layer differ from organ to organ?

> **ⓘ** *safety first*
> Observe the usual precautions when using the microscope and prepared slides.  •

☐ 2  Obtain a prepared slide of a sectioned stomach wall and examine it with your microscope. Identify the four layers of the wall. Examine the mucosa. Notice that it is composed of simple columnar epithelium that forms deep **gastric pits.** They are lined with **surface mucous cells,** which you may recognize as goblet cells. Follow a pit's lumen to the base, or **gastric gland.** You may see more goblet cells called **mucous neck cells.** Some of the very darkly stained cells are **chief cells.** The very pale cells are **parietal cells.** What is the function of each of these cells?
☐ 3  Obtain a prepared slide of a small intestine section. Identify the four layers of the wall. In a true cross-section, you cannot tell that the simple columnar epithelium forms large **circular folds,** but you may have noticed them in the model. Each circular fold has fingerlike projections called **villi** that are visible in your specimen.
☐ 4  Obtain a prepared slide of a salivary gland section. Identify the mucus-producing glandular cells by their characteristic light appearance that is produced by the transparent nature of the mucus that fills them. Try to find the cuboidal duct cells that form rings in a cross-sectioned specimen.

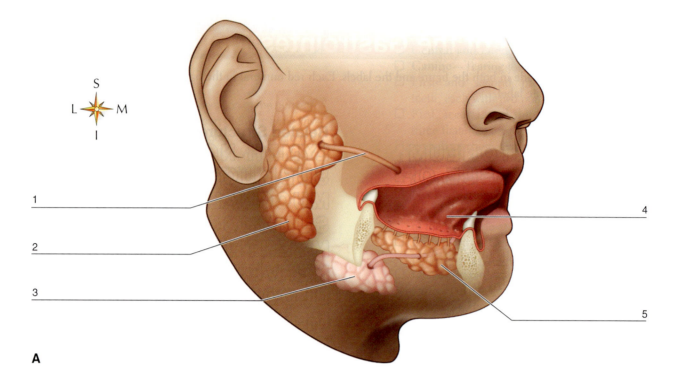

**A**

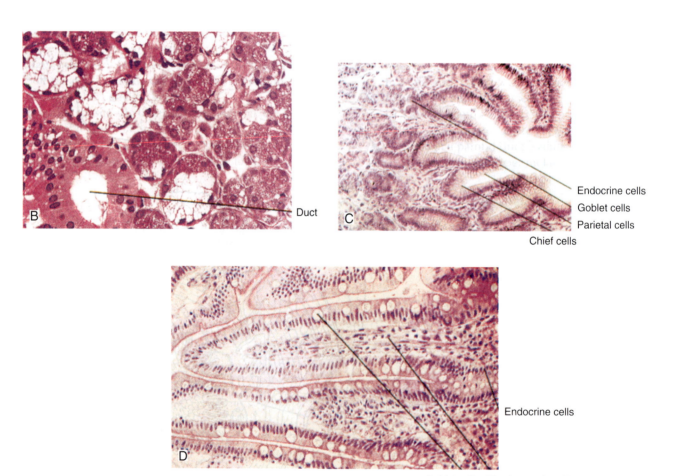

Duct

Endocrine cells
Goblet cells
Parietal cells
Chief cells

Endocrine cells

Goblet cell    Blood vessel and lacteal

**FIGURE 44-2  A,** Identify the features of the mouth indicated on the lines provided and on the blanks in the Lab Report at the end of this exercise. **B,** Salivary gland tissue. **C,** Epithelium of the stomach. **D,** Epithelium of the small intestine.

## Coloring Exercises: Mouth and Teeth

Use colored pens or pencils to shade in both the figure and the labels. Each red numeral in the figure corresponds to a matching red numeral following the appropriate label.

### Mouth

LIPS 1
BUCCINATOR MUSCLE 2
TONGUE 3
HARD PALATE 4
SOFT PALATE 5
UVULA 6
PALATINE TONSILS 7
OROPHARYNX 8

### Teeth

**Tooth Types**

CENTRAL INCISOR 9
LATERAL INCISOR 10
CANINE 11
FIRST PREMOLAR 12
SECOND PREMOLAR 13
FIRST MOLAR 14
SECOND MOLAR 15
THIRD MOLAR 16

**Tooth Features**

CROWN 17
NECK 18
ROOT 19
PULP CAVITY 20
PULP 21
DENTIN 22
ENAMEL 23
GINGIVA 24
PERIODONTAL
   MEMBRANE 25
PERIODONTAL
   LIGAMENTS 26
ALVEOLAR RIDGE 27

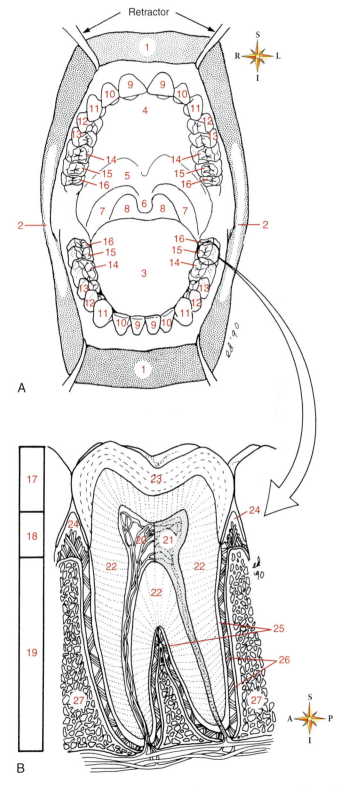

FIGURE 44-3 **A,** Structure of the oral cavity. **B,** Structure of a tooth.

# clinical application: *fluoroscopy*

**Fluoroscopy** is a type of radiographic imaging that uses a fluorescent screen rather than film. X-rays pass through the patient and are absorbed by the screen's coating. The energy is reemitted by the coating as light energy. Dense tissues cast a dark shadow on the screen. Therefore continuous radiation allows a continuous moving image in a moving patient.

Fluoroscopy with the contrast medium **barium sulfate** is used to visualize portions of the digestive tract. A *barium swallow*, or **upper GI series,** images the stomach and portions of the esophagus and small intestine. A *barium enema* (**lower GI series**) images the rectum and colon. A **hernia** of the digestive tract, where a portion of the tract is pushed out of a weak spot in the abdominopelvic wall,

can be detected this way. **Ulcers,** holes in the digestive wall, can also be assessed in this manner.

Because fluoroscopy allows observation of the body in motion, it can be used to see the muscular movements of the digestive tract as the barium passes through it. A person can push on the abdomen from the outside to see how digestive organs withstand pressure, similar to compression of abdominal muscles during heavy lifting.

**Figures 44-5, *A*** and ***B*,** are regular x-ray photographs taken of an upper and lower GI series. A fluoroscope image would be reversed (black on white, not the white on black seen here). Try to identify the labeled parts on the lines provided and on the blanks in the Lab Report at the end of this exercise.

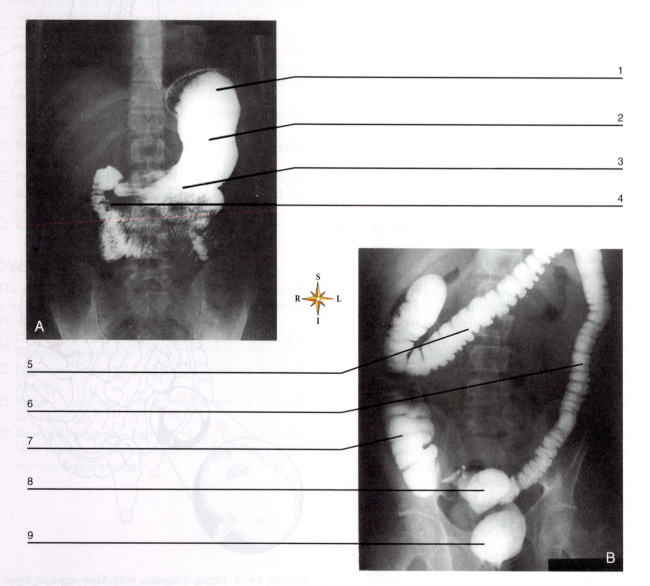

**FIGURE 44-5** Radiographs of **A,** a barium swallow, or upper GI and, **B,** a barium enema, or lower GI. NOTE: Fluoroscope images appear black on white, the reverse of the white-on-black images shown here.

Name: _____   Date: _____   Section: _____

## LAB REPORT 44
# Digestive Structures

Specimen: *stomach wall c.s.*
Total Magnification: _____

Specimen: *small intestine wall c.s.*
Total Magnification: _____

Specimen: *salivary gland c.s.*
Total Magnification: _____

| STRUCTURE | MODEL | FUNCTION(S) |
|---|---|---|
| Oral cavity | ☐ | |
| Buccinator muscle | ☐ | |
| Tongue | ☐ | |
| Palate: hard, soft | ☐ | |
| Uvula | ☐ | |
| Palatine tonsils | ☐ | |
| Incisors: central, lateral | ☐ | |
| Canine teeth | ☐ | |
| Premolars: first, second | ☐ | |
| Molars: first, second, third | ☐ | |
| Parotid glands | ☐ | |
| Submandibular glands | ☐ | |
| Sublingual glands | ☐ | |
| Pharynx | ☐ | |
| Esophagus | ☐ | |
| Cardiac opening, region | ☐ | |
| Fundus | ☐ | |
| Body of stomach | ☐ | |
| Stomach curvatures: greater, lesser | ☐ | |
| Pyloric opening, region | ☐ | |
| Rugae | ☐ | |
| Duodenum | ☐ | |
| Jejunum | ☐ | |
| Ileum | ☐ | |
| Circular folds | ☐ | |
| Ileocecal junction: valve, sphincter | ☐ | |

*Continued*

| STRUCTURE | MODEL | FUNCTION(S) |
|---|---|---|
| Liver | ☐ | |
| Common bile duct | ☐ | |
| Common hepatic duct | ☐ | |
| Cystic duct | ☐ | |
| Pancreas | ☐ | |
| Pancreatic duct | ☐ | |
| Cecum | ☐ | |
| Appendix | ☐ | |
| Ascending colon | ☐ | |
| Transverse colon | ☐ | |
| Descending colon | ☐ | |
| Sigmoid colon | ☐ | |
| Rectum | ☐ | |
| Anal canal, anus | ☐ | |
| Lesser omentum | ☐ | |
| Greater omentum | ☐ | |

## Figure 44-2 (from p. 426)

1. _____
2. _____
3. _____
4. _____
5. _____

## Figure 44-5 (from p. 430)

1. _____
2. _____
3. _____
4. _____
5. _____
6. _____
7. _____
8. _____
9. _____

## Fill-in

1. _____
2. _____
3. _____
4. _____
5. _____
6. _____
7. _____

## Fill-in (complete each statement with the correct term)

1. The stomach's mucosa forms large folds called _?_.

2. The two ducts that exit the liver and join to form the common hepatic duct are called _?_ ducts.

3. The colon is divided into _?_ sections.

4. The_?_ duct empties into the duodenum.

5. A substance called _?_ covers the dentin of the tooth's crown.

6. The _?_ is the portion of the pharynx posterior to the mouth cavity.

7. The _?_ sphincter prevents stomach contents from flowing back into the esophagus.

# Dissection: Digestive System

In this exercise, you continue your study of digestive anatomy by dissecting a whole preserved specimen.

Activity A provides directions for studying the digestive anatomy of the cat. Activity B is an alternate activity, providing directions for studying the digestive anatomy of the fetal pig.

## BEFORE YOU BEGIN

☐ Read the appropriate chapter in your textbook.
☐ Set your learning goals. When you finish this exercise, you should be able to:
  ☐ dissect the digestive anatomy of a preserved cat or fetal pig
  ☐ identify the primary organs of the digestive tract in a dissected mammalian specimen
  ☐ identify the accessory organs of digestion in a dissected mammalian specimen
☐ Prepare your materials:
  ☐ preserved cat or fetal pig
  ☐ dissection tools and trays
  ☐ protective gear: gloves, safety eyewear, and apron
  ☐ storage container (if specimen is to be reused)
☐ Read the directions and safety tips for this exercise *carefully* before starting any procedure.

> **! safety first**
> Observe the usual precautions when working with a preserved specimen. Heed the safety advice accompanying preservatives used with your specimen. Use protective gloves while handling your specimen. Avoid injury with dissection tools. Use safety goggles to avoid injury during dissections. Dispose of or store your specimen as instructed. •

## *Study* TIPS

As recommended in the other dissections, as you complete this dissection, use a diagram or illustration to compare your specimen with that of the human digestive tract. For reference, use your lab manual, textbook, or the *A&P Survival Guide*. Use the camera on your cell phone to take photos of the specimen and use them later as a resource.

## A. DIGESTIVE ANATOMY OF THE CAT

This activity challenges you to identify the major digestive structures of the cat. Assuming you have already removed the cat's skin (Lab Exercise 10—The Skin) and opened the abdominopelvic cavity (Lab Exercise 40—Dissection: Cardiovascular and Lymphatic Systems), you simply have to move the viscera out of the way (without cutting or removing them) to see the structures indicated in this activity.

> **Hint** ▶ **Figure 45-1** illustrates many of the structures of the cat listed in this activity. •

☐ 1 Identify the **peritoneum,** the serous membrane that lines the wall of the abdominopelvic cavity (*parietal peritoneum*) and covers the organs within the cavity (*visceral peritoneum*). Identify these extensions or folds of the peritoneum:
  ☐ **Greater omentum**
  ☐ **Lesser omentum**
  ☐ **Mesentery**

> **Hint** ▶ The *greater omentum* is a large, fatty fold of the peritoneum that lies like an apron over the ventral surfaces of the abdominopelvic organs. You must lift or cut this "apron" out of the way to see most of the digestive organs described in the following steps. •

☐ 2 Identify the **liver** with its *left lobes* and *right lobes*. Locate the **gallbladder** at the posterior surface of the right lobes of the liver. Trace the **cystic duct** from the gallbladder and the **hepatic ducts** from the lobes of the liver. Note where they join to form the **common bile duct**.

☐ 3 Locate the curved **stomach**. Identify these regions of the stomach:
  ☐ **Fundus**
  ☐ **Body**
  ☐ **Pylorus**
  ☐ **Rugae**—Cut the stomach open to identify these mucosal folds.

☐ 4 Identify the **pancreas**. It is composed of two lobes of glandular tissue in the cat, one that is embedded in the *mesentery* along the duodenum and one that extends along the edge of the stomach.

☐ 5 The **small intestine** extends posteriorly from the pylorus of the stomach. Identify these portions of the small intestine:
  ☐ **Duodenum**
  ☐ **Jejunum**
  ☐ **Ileum**

☐ 6 The **large intestine** attaches to the distal end of the ileum of the small intestine. Locate these portions of the large intestine:
  ☐ **Cecum**
  ☐ **Ascending colon**
  ☐ **Transverse colon**
  ☐ **Descending colon**
  ☐ **Rectum**
  ☐ **Anus**—Opening to the exterior

# B. DIGESTIVE ANATOMY OF THE FETAL PIG

This activity challenges you to identify the major digestive structures of the fetal pig. Assuming you have already removed the animal's skin (Lab Exercise 10—The Skin) and opened the abdominopelvic cavity (Lab Exercise 40—Dissection: Cardiovascular and Lymphatic Systems), you simply have to move the viscera out of the way (without cutting or removing them) to see the structures indicated in this activity.

> **Hint** ▸ **Figure 45-2** illustrates many of the structures of the fetal pig listed in this activity.  •

☐ 1  Identify the **peritoneum**, the serous membrane that lines the wall of the abdominopelvic cavity (*parietal peritoneum*) and covers the organs within the cavity (*visceral peritoneum*). Identify these extensions or folds of the peritoneum:
  ☐ **Greater omentum**
  ☐ **Lesser omentum**
  ☐ **Mesentery**

> **Hint** ▸ The *greater omentum* is a large, fatty fold of the peritoneum that lies like an apron over the ventral surfaces of the abdominopelvic organs. You must lift or cut this "apron" out of the way to see most of the digestive organs described in the following steps.  •

☐ 2  Identify the **liver** with its *left lobes* and *right lobes*. Locate the **gallbladder** at the posterior surface of the right lobes of the liver. Trace the **cystic duct** from the gallbladder and the **hepatic ducts** from the lobes of the liver. Note where they join to form the **common bile duct**.

☐ 3  Locate the curved **stomach**. Identify these regions of the stomach:
  ☐ **Fundus**
  ☐ **Body**
  ☐ **Pylorus**
  ☐ **Rugae**—Cut the stomach open to identify these mucosal folds.

☐ 4  Identify the **pancreas**, a mass of glandular tissue that lies within the curve of the duodenum of the small intestine.

☐ 5  The **small intestine** extends posteriorly from the pylorus of the stomach. Identify these portions of the small intestine:
  ☐ **Duodenum**
  ☐ **Jejunum**
  ☐ **Ileum**

☐ 6  The **large intestine** attaches to the distal end of the ileum of the small intestine. There are two main portions of the large intestine.
  ☐ **Spiral colon**—Proximal portion
  ☐ **Rectum**—Distal portion leading to the opening to the exterior, the **anus**

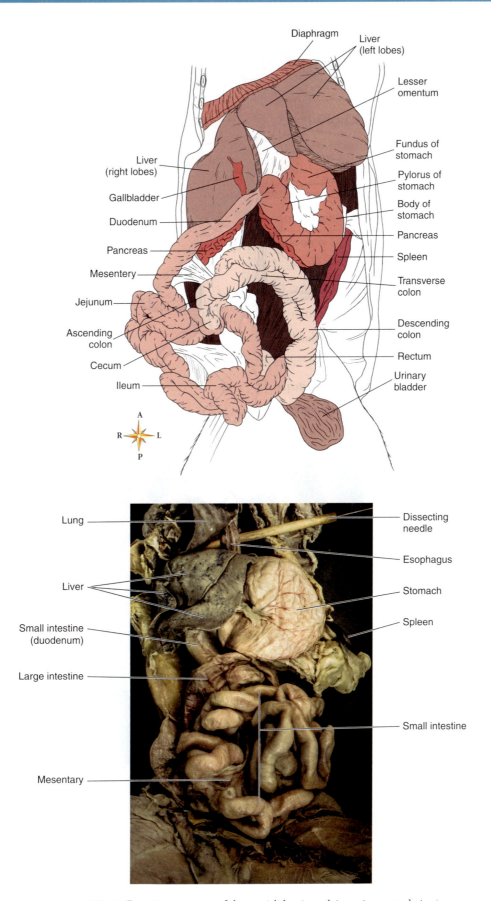

Diaphragm

Liver
(left lobes)

Lesser
omentum

Liver
(right lobes)

Fundus of
stomach

Pylorus of
stomach

Gallbladder

Body of
stomach

Duodenum

Pancreas

Pancreas

Spleen

Mesentery

Transverse
colon

Jejunum

Descending
colon

Ascending
colon

Rectum

Cecum

Urinary
bladder

Ileum

A
R — L
P

Lung

Dissecting
needle

Esophagus

Liver

Stomach

Small intestine
(duodenum)

Spleen

Large intestine

Small intestine

Mesentary

**FIGURE 45-1** Digestive anatomy of the cat (abdominopelvic cavity, ventral view).

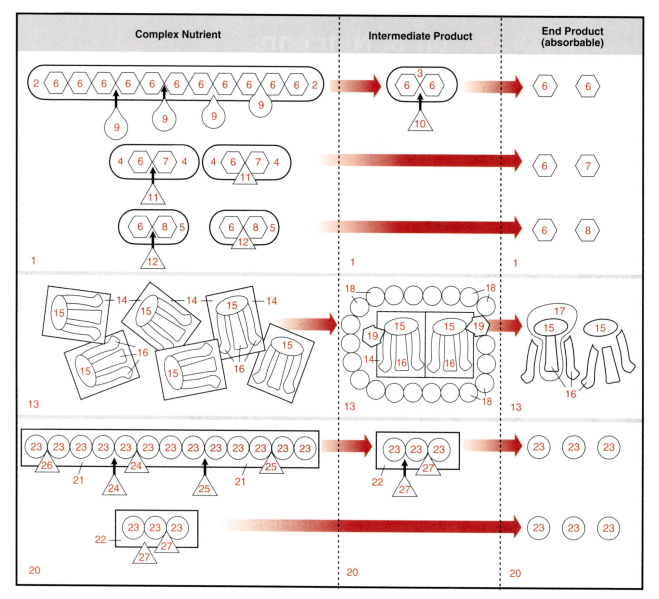

FIGURE 46-2 Digestive processes.

In this exercise, you will use an extract of pancreatic secretions called *pancreatin*. Pancreatin contains pancreatic amylase, as well as pancreatic **lipase** and pancreatic **protease**.

☐ 1  First, practice the tests used to analyze solutions for the presence of starch and its sugar products.

    ☐ **Lugol test**—The Lugol test uses *Lugol reagent* (an iodine solution) to test for the presence of starch. If Lugol iodine turns from its normal orange color to a deep blue-black, starch is present. Mark a test tube **1a** and put in 2 mL (two full droppers) of starch solution. Add a few drops of Lugol reagent. Notice the color change. Anytime you see this result for Lugol test, you have a positive result (meaning starch *is* present). Mark another tube **1b** and put in 2 mL of distilled water and a few drops of Lugol reagent. The solution should be a shade of orange, which is a negative result (meaning starch *is not* present). Record your results in **Figure 46-3**, and then, interpret your results in the Lab Report at the end of this exercise. Also, save the tubes and their contents for later reference.

    ☐ **Benedict test**—The Benedict test uses Benedict reagent to determine whether a solution contains certain types of sugar. Mark a test tube **2a** and add 2 mL of maltose solution. To this, add 2 mL of **Benedict reagent**. In a tube marked **2b** put 2 mL of distilled water and 2 mL of Benedict reagent. Place each tube in a beaker of boiling water on a hot plate for 3 minutes (see **Figure 46-3, B**). At the end of 3 minutes, remove the tubes and place them in your rack. Tube **2a** should show a positive result, or a change from blue to another color, meaning sugar is present. Tube **2b** is an example of a negative result, with the contents remaining blue. Record your results in **Figure 46-3, A**, and then, interpret your results in the Lab Report at the end of this exercise. Also, save these tubes for later reference.

☐ 2  Mark eight test tubes: **3a, 3b, 4a, 4b, 5a, 5b, 6a,** and **6b**. To each add 2 mL of starch. Next, simultaneously add a small amount of pancreatin (just enough powder to cover the end of the spatula) to each tube. Immediately:

☐ Place tubes **3a** and **3b** in an ice water bath (0°C to 1°C)

☐ Place tubes **4a** and **4b** in a rack at room temperature (20°C to 25°C)

☐ Place tubes **5a** and **5b** in a warm water bath set at body temperature (37°C)

☐ Place tubes **6a** and **6b** in a boiling water bath (100°C)

☐ Allow each tube to stand in its respective bath for 30 minutes. Remove tubes **6a** and **6b** from the boiling water earlier if they decrease to 0.5 mL or less as the result of vaporization.

☐ 3  At the end of 30 minutes, place all test tubes in the rack. Test all the **a** tubes for starch by adding a few drops of Lugol reagent to each. Test all the **b** tubes for sugar by adding 2 mL of Benedict reagent to each and boiling them for 3 minutes. Record your results in **Figure 46-3, A**, and then, interpret your results in the Lab Report at the end of this exercise. Did starch digestion occur in any of the tubes? Which ones? How do you know? Did temperature affect digestion? How? Can you explain this result?

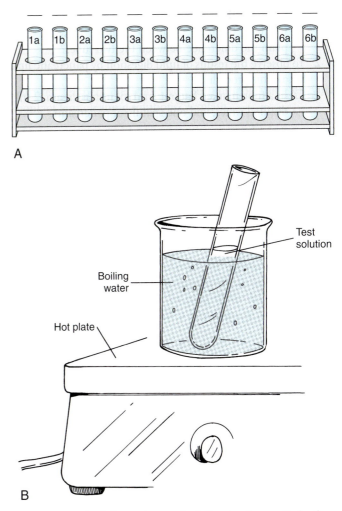

A

B

**FIGURE 46-3  A,** Color each test tube indicating the results for the Lugol test or Benedict test. Also, above each test tube indicate whether the test was positive (+) or negative (−). **B,** A boiling water bath, as used in Benedict test for the presence of sugars.

Urine formed by the kidney is conducted through the collecting ducts of the renal pyramids to the calyces. The calyces fuse to form the large renal pyramid. As urine is conducted out of the kidney through the hilum, the urinary channel narrows to form the ureter.

☐  **4** Identify these features of the renal arterial supply and capillaries **(Figure 47-2)**:

☐  **Renal artery**—A branch of the abdominal aorta that enters the hilum and extends through the renal sinus

☐  **Interlobar arteries**—Branches of the renal arteries that extend outward through the tissue between the pyramids

☐  **Arcuate arteries**—Branches of the interlobar arteries that turn to extend between the cortex and medulla

☐  **Interlobular arteries**—Branches of the arcuate arteries that extend outward into the cortex

☐  **Afferent arterioles**—Small arteries that arise from branches of interlobular arteries, each one extending to a glomerulus

☐  **Glomerular capillaries**—Capillaries that arise from an afferent arteriole and form a ball or small capillary bed called a **glomerulus**

☐  **Efferent arterioles**—Each extending from a glomerulus

☐  **Peritubular capillaries**—Small capillary beds that each arise from an efferent arteriole (a network of peritubular capillaries surrounds each *nephron*, the tubular, microscopic unit of the kidney)

☐  **5** The venous network of the kidney parallels the arterial network, as you may expect (see **Figure 47-2**):

☐  **Interlobular veins**

☐  **Arcuate veins**

☐  **Interlobar veins**

☐  **Renal veins**—Blood vessels that extend through the renal sinus, out the hilum, to drain into the inferior vena cava

## C. THE SHEEP KIDNEY

The sheep kidney is similar to the human kidney and makes an ideal specimen for study **(Figure 47-3)**.

> **! safety first**
>
> Observe the usual precautions when dissecting preserved specimens. Be sure to follow the safety advice that accompanies the preservative used in your specimen. Avoid injury with the dissection tools. Use safety goggles to avoid injury during dissections.  •

☐  **1** Examine the external aspect of your specimen. Identify as many parts of the kidney as you can. Refer to Activities A and B if needed. You may have to remove some of the renal fat pad.

☐  **2** Use a long knife or scalpel to cut a section, dividing the kidney into roughly equal dorsal and ventral portions. Try to identify as many features as you can.

> **Hint** ▶ Double-injected specimens have red latex injected into the arteries and blue latex injected into the veins. Triple-injected specimens also have yellow latex injected into the urinary channels and tubules. You may be able to see some of the finer detail visible in an injected specimen by using a hand lens or a dissection microscope.  •

Na

## Coloring Exercises: Urinary System

Use colored pens or pencils to shade in both the figure and the labels. Each red numeral in the figure corresponds to a matching red numeral following the appropriate label.

KIDNEY 1
URETER 2
URINARY BLADDER 3
URETHRA 4
RENAL CAPSULE 5
HILUM 6
RENAL SINUS 7
RENAL PELVIS 8
CALYCES 9
RENAL CORTEX 10
RENAL MEDULLA 11
RENAL PYRAMIDS 12
RENAL ARTERY 13
RENAL VEIN 14
NEPHRON 15

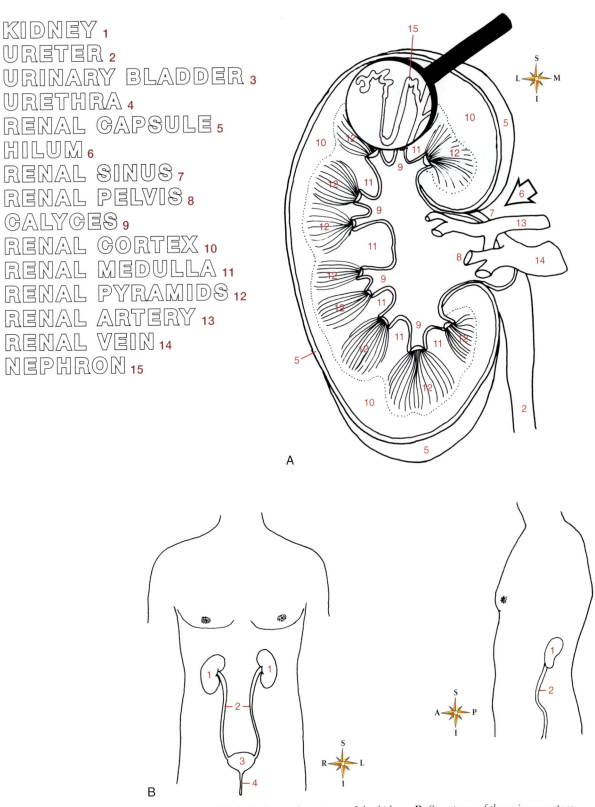

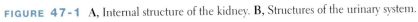

**FIGURE 47-1** **A,** Internal structure of the kidney. **B,** Structures of the urinary system.

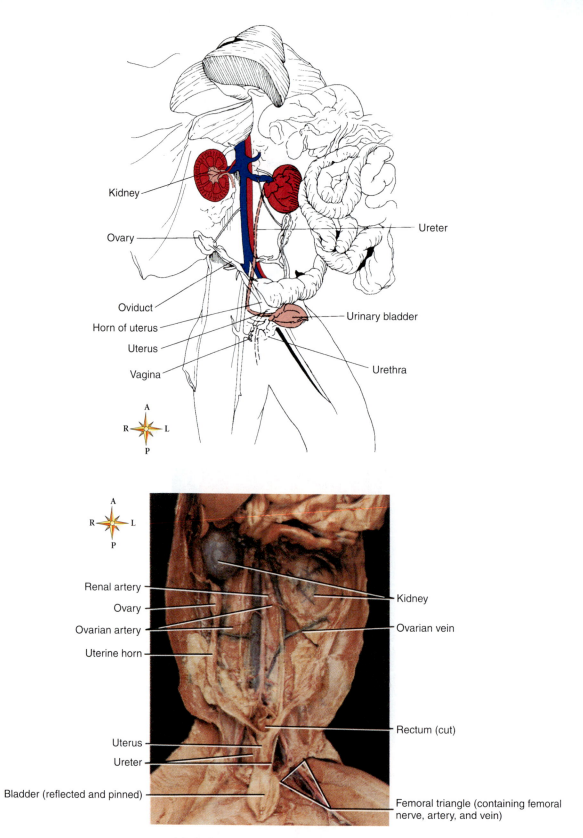

**FIGURE 48-2** Urinary anatomy of the female cat. Ventral view.

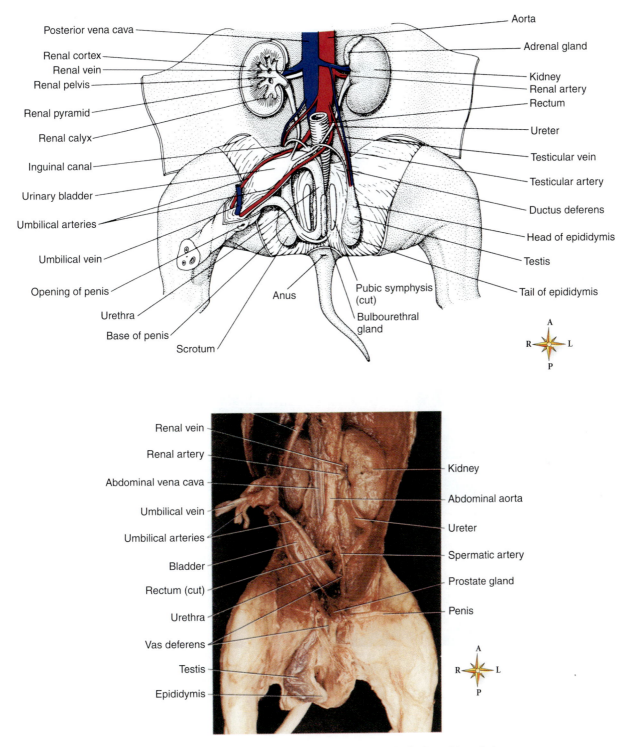

Posterior vena cava

Renal cortex
Renal vein
Renal pelvis

Renal pyramid

Renal calyx

Inguinal canal

Urinary bladder

Umbilical arteries

Umbilical vein

Opening of penis

Urethra

Base of penis

Scrotum

Anus

Pubic symphysis (cut)

Bulbourethral gland

Aorta

Adrenal gland

Kidney
Renal artery

Rectum

Ureter

Testicular vein

Testicular artery

Ductus deferens

Head of epididymis

Testis

Tail of epididymis

A
R — L
P

Renal vein

Renal artery

Abdominal vena cava

Umbilical vein

Umbilical arteries

Bladder

Rectum (cut)

Urethra

Vas deferens

Testis

Epididymis

Kidney

Abdominal aorta

Ureter

Spermatic artery

Prostate gland

Penis

A
R — L
P

**FIGURE 48-3** Urinary anatomy of the male fetal pig. Ventral view.

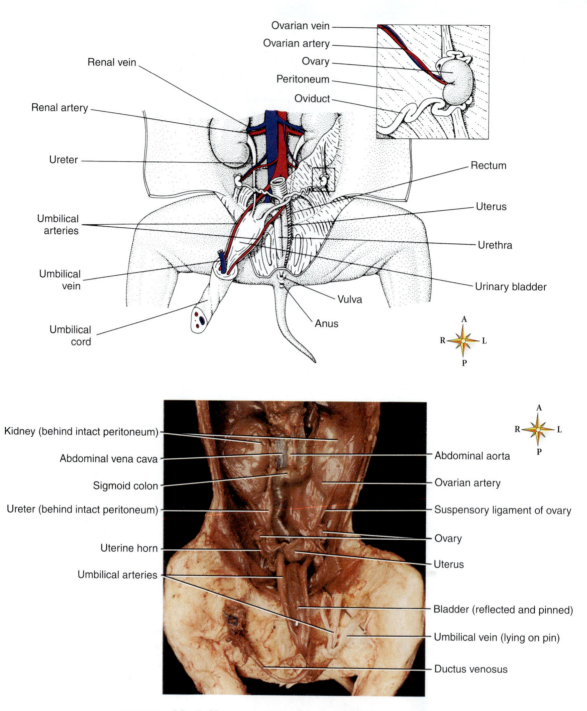

Ovarian vein
Ovarian artery
Ovary
Peritoneum
Oviduct

Renal vein
Renal artery
Ureter
Umbilical arteries
Umbilical vein
Umbilical cord

Rectum
Uterus
Urethra
Urinary bladder
Vulva
Anus

Kidney (behind intact peritoneum)
Abdominal vena cava
Sigmoid colon
Ureter (behind intact peritoneum)
Uterine horn
Umbilical arteries

Abdominal aorta
Ovarian artery
Suspensory ligament of ovary
Ovary
Uterus
Bladder (reflected and pinned)
Umbilical vein (lying on pin)
Ductus venosus

**FIGURE 48-4**   Urinary anatomy of the female fetal pig. Ventral view.

Name: _____    Date: _____    Section: _____

**LAB REPORT 48**
# Dissection: Urinary System

## A. Cat Dissection Checklist

- ☐ parietal peritoneum
- ☐ renal fat pad
- ☐ adrenal gland
- ☐ ureter
- ☐ renal artery
- ☐ renal vein
- ☐ kidney (left and right)
  - ☐ hilum
  - ☐ renal capsule
  - ☐ renal cortex
  - ☐ renal medulla
  - ☐ renal pyramid
  - ☐ renal papilla
  - ☐ renal pelvis
- ☐ urethra (male)
- ☐ urethra (female)

## B. Fetal Pig Dissection Checklist

- ☐ parietal peritoneum
- ☐ renal fat pad
- ☐ adrenal gland
- ☐ ureter
- ☐ renal artery
- ☐ renal vein
- ☐ kidney (left and right)
  - ☐ hilum
  - ☐ renal capsule
  - ☐ renal cortex
  - ☐ renal medulla
  - ☐ renal pyramid
  - ☐ renal papilla
  - ☐ renal pelvis
- ☐ urethra (male)
- ☐ urethra (female)

# Urinalysis

The examination of urine and its contents is **urinalysis**. Urinalysis can be very comprehensive, testing for dozens of different physical and chemical characteristics. On the other hand, some clinical situations call for urinalysis that tests only one or two urine characteristics. For example, the pregnancy test performed in Lab Exercise 33—Hormones was a type of urinalysis that determined the presence or concentration of one particular hormone. In this exercise, you will perform some of the more common clinical urinalysis tests, using the most current methods.

## BEFORE YOU BEGIN

- [ ] Read the appropriate chapter in your textbook.
- [ ] Set your learning goals. When you finish this exercise, you should be able to:
  - [ ] state the normal characteristics of freshly voided urine
  - [ ] demonstrate the use of dip-and-read clinical test strips for urinalysis
  - [ ] explain the significance of common urine tests
  - [ ] prepare and examine a stained urine sediment slide
  - [ ] evaluate urinalysis results
- [ ] Prepare your materials:
  - [ ] fresh urine specimen using a packaged substitute in a disposable container
  - [ ] dip-and-read multiple test strips (Multistix 10 SG are preferred)
  - [ ] urine centrifuge and tapered tubes
  - [ ] disposable 1-ml droppers
  - [ ] Sedi-Stain urine sediment stain
  - [ ] paper towels and wipes
  - [ ] microscope slides and coverslips
  - [ ] compound light microscope
  - [ ] full-color urine sediment chart (optional)
  - [ ] BIOHAZARD container
  - [ ] unknown urine additives (optional)
  - [ ] protective gear: gloves, safety eyewear, and apron
- [ ] Read the directions and safety tips for this exercise *carefully* before starting any procedure.

## Study TIPS

Urine is a filtrate of plasma. Think of a urinalysis as a "fluid biopsy" of the kidney. A urinalysis is used to evaluate or monitor body homeostasis and detect many metabolic diseases. As you perform this lab activity, understand how certain diseases may affect the characteristics of urine. Table 31-2 in your textbook describes the normal and abnormal characteristics of urine.

## A. EXAMINATION OF URINE

Urine reflects the overall status of extracellular fluid because it is derived from blood, and its contents have been adjusted to some extent on the basis of homeostatic balance. Additionally, the health of the kidneys and urinary tract in particular affects the characteristics of urine. Observe the urinalysis tests described or perform them yourself on a packaged substitute provided by your instructor.

> **safety first**
>
> Use artificial or prepackaged sterile human urine. You should still protect yourself and others from contamination by wearing protective lab apparel, gloves, and eyewear. Disinfect all surfaces that have, or *could have,* come into contact with urine. Put all disposable urine containers, droppers, towels, wipes, slides, and coverslips in a BIOHAZARD container *immediately* after use. Handle the sample as if it was nonsterile, freshly collected urine in an effort to practice good laboratory technique. •

- [ ] 1 Examine these physical characteristics of urine after placing some of the sample in a transparent container:
  - [ ] **Transparency**—Normal urine is clear to slightly cloudy (especially after standing). Cloudy urine may contain fat globules, epithelial cells, mucus, microbes, or chemicals.
  - [ ] **Color**—Normal urine is amber, straw, or transparent yellow, resulting from the presence of bile pigments called **urochromes** or **urobilins** (products of *bilirubin* metabolism). Yellow-brown to greenish urine may occur when a high concentration of bile pigments is present. Urine that is red to dark brown may have blood present. These or other abnormal colors may also result from the presence of food pigments (e.g., carotene), drugs, or other chemicals.
- [ ] 2 Many characteristics of urine are determined by the use of paper that is impregnated with test reagents. The paper is dipped into a sample, and a color change indicates the presence (and sometimes the concentration) of a particular substance. You may have used pH test paper in your aquarium or in another lab course. In this step, you will use a plastic strip on which 9 or 10 test papers have been placed. Each paper is impregnated with different reagents. Shake or stir the sample so that any sediment on the bottom of the container becomes suspended. With gloved hands, dip the papered end of one test strip into your sample. Lift it out and tap the excess urine onto the inside rim of the container. You may need to transfer some urine to a test tube so that you can immerse all the paper pads.

Read the results by comparing the test papers with standard charts as instructed. Usually, certain tests *must* be read at certain times. A 10-test strip tests for these urine components:

> **Hint** ▸ Carefully read the instructions provided by the distributor of your test strips. Follow the instructions exactly, especially those dealing with times at which test strips are to be read.  •

- ☐ **Leukocytes**—The occasional presence of white blood cells (WBCs) is normal, but values increase in urinary infections.
- ☐ **Nitrite**—A positive nitrite result indicates the presence of large amounts of bacteria, as in an infection. This test is useful if one has a cloudy urine sample but no microscope to determine whether bacteria are causing the cloudiness.
- ☐ **Urobilin**—A derivative of bilirubin, high levels may indicate excessive red blood cell (RBC) destruction or liver disease.
- ☐ **Protein**—Specifically *albumin*, a small protein molecule that is normally absent or present only in trace amounts; higher levels may indicate hypertension or kidney disease. Detectable levels are normal just after exercise.
- ☐ **pH**—The relative $H^+$ concentration, pH is a determinant of acidity. Normal urine pH is 4.6 to 8. Values are lower (more acid) in acidosis, starvation, and dehydration. pH is higher (more alkaline) in urinary infections and alkalosis.
- ☐ **Occult blood** (RBCs or free hemoglobin)—*Occult* means hidden. This strip tests for small amounts of hemoglobin that do not discolor the urine but are still clinically significant. Normally not present, hemoglobin may indicate kidney infection or the presence of stones in the kidney, ureter, or bladder.
- ☐ **Specific gravity**—This is the ratio of urine density to water density. If urine is pure water, the specific gravity is 1.000. Normal urine is 1.001 to 1.030. Lower values may indicate kidney disease. Higher values indicate high solute concentration and may occur during dehydration or diabetes mellitus. (This test is not present in 9-test strips.) An alternate method for determining specific gravity is to float a *hydrometer* in a cylinder containing urine, as in **Figure 49-1**. Read the scale on the hydrometer at the point that it meets the surface of the urine.
- ☐ **Ketone**—A by-product of fat metabolism, it may be present during fasting, diabetes mellitus, or a low-carbohydrate diet.
- ☐ **Bilirubin**—Normally not present, or only trace amounts present, this product of RBC destruction in the liver may indicate liver disease or bile tract obstruction if present in the urine.
- ☐ **Glucose**—Normally, there is no glucose in the urine. Trace amounts may be present after a meal high in carbohydrates. Continued high levels in the urine may indicate diabetes mellitus or pituitary problems.

**FIGURE 49-1** The specific gravity of urine may be determined by floating a hydrometer in urine at room temperature.

> **⚠ safety first**
> Do not forget to observe the usual precautions when using the microscope and stained, wet-mount slides.  •

☐ 3 Observe the components of urine sediment by performing this procedure:
   - ☐ Shake or stir the urine sample to suspend any sediment that has settled.
   - ☐ Transfer some of the urine to a tapered centrifuge tube.
   - ☐ After balancing and securing the centrifuge as your instructor demonstrates, spin the sample for about 5 minutes.
   - ☐ Without disturbing the sediment that has collected in the tip of the tube, squeeze the bulb of a disposable dropper and gently lower it into the urine. When the dropper reaches the bottom, release the bulb *slightly* to collect only a few drops of sediment-containing urine.
   - ☐ Quickly place a drop of the sample on a clean microscope slide. Add one drop of Sedi-Stain and cover it with a coverslip. Remove excess fluid from the edges with a lab wipe and immediately discard the wipe in the BIOHAZARD container.
   - ☐ Examine the specimen under both low and high power. Move from field to field as you identify different types of sediment.

CELLS

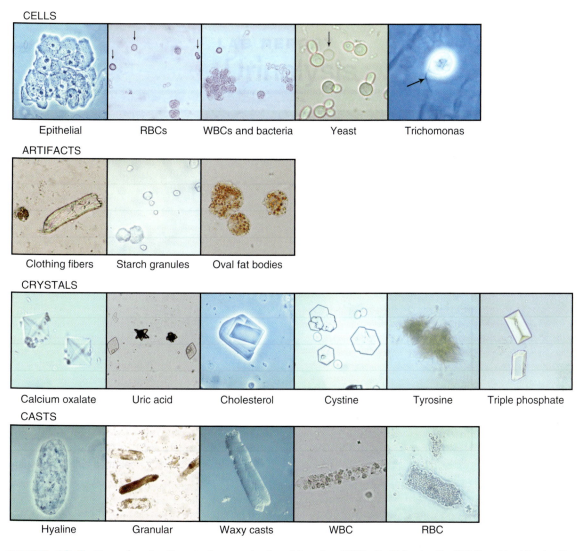

Epithelial  RBCs  WBCs and bacteria  Yeast  Trichomonas

ARTIFACTS

Clothing fibers  Starch granules  Oval fat bodies

CRYSTALS

Calcium oxalate  Uric acid  Cholesterol  Cystine  Tyrosine  Triple phosphate

CASTS

Hyaline  Granular  Waxy casts  WBC  RBC

**FIGURE 49-2** Examples of sediments that may be found in urine. *RBCs*, Red blood cells; *WBCs*, white blood cells.

☐ 4 Your instructor may have a chart of sediment types for you to use in identifying urine solids. **Figure 49-2** presents a minichart of common urine sediment components:

☐ **Cells**—Epithelial cells from the urinary tract lining, blood cells from injury or infection sites, or infectious microbes may be present in sediment. If more than trace amounts of blood or microbial cells are present, a urinary problem is indicated.

☐ **Artifacts**—Artifacts are materials that have accidentally gotten into the sample. These include fabric fibers from underwear, powder used on the skin near the urethral opening, or skin oil droplets.

☐ **Crystals**—Very tiny crystals of normal urine components or drugs may be visible under high power. Small amounts of these crystals may be normal. A large number of crystals are seen in urinary *retention*, the inability to void urine from the bladder. Large masses of crystals are called *stones*, or *calculi*.

☐ **Casts**—Casts are chunks of material that have hardened somewhere in the urinary channel and sloughed off into the urine. They may be roughly cylindrical masses of cells, granules, or other substances.

☐ 5 Your instructor may opt to offer unknown urine additives. Add the additive to your sample as instructed and test the sample as if it were a new urine specimen. The additive will have altered the sample's characteristics in a way that mimics an abnormal condition. Your instructor may choose instead to give you a new sample, one that is abnormal in some respect. Test the new sample in the same way you tested the first sample. Report your results in Lab Report at the end of this exercise.

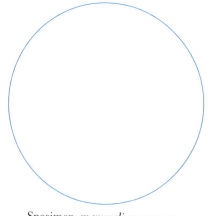

Specimen: *mammalian ovary c.s.*
Total Magnification: _____

**Figure 51-2** (from p. 484)

1. _____
2. _____
3. _____
4. _____
5. _____
6. _____
7. _____
8. _____
9. _____
10. _____

**Fill-in**

1. _____
2. _____
3. _____
4. _____
5. _____
6. _____
7. _____
8. _____
9. _____
10. _____
11. _____
12. _____

**Fill-in** (complete each statement with the correct term)

1. The fluid-filled space within a follicle is called the _?_.
2. The tissue surrounding a mature follicle is called the _?_.
3. A follicle secretes the hormone _?_.
4. The external genitals of the female are known by the term _?_.
5. The _?_ secrete lubricating fluid into the vestibule.
6. The release of an oocyte from a follicle is termed _?_.
7. The normal site of fertilization of an egg is in the _?_.
8. The narrow portion, or neck, of the uterus is called the _?_.
9. The oocyte in a secondary follicle is embedded in the _?_ mass.
10. The thin membrane covering all or part of the vaginal opening is called the _?_.
11. The _?_ is a structure of the ovary that secretes progesterone.
12. The birth canal is also called the _?_.

# Dissection: Reproductive Systems

In this exercise, you continue your study of reproductive anatomy by dissecting a whole preserved specimen.

Activity A provides directions for studying the reproductive anatomy of the male cat. Activity B provides directions for studying the reproductive anatomy of the female cat. Activities C and D are alternate activities, providing directions for studying the immature reproductive anatomy of the male and female fetal pig.

## BEFORE YOU BEGIN

☐ Read the appropriate chapter in your textbook.
☐ Set your learning goals. When you finish this exercise, you should be able to:
  ☐ dissect the reproductive anatomy of a preserved cat or fetal pig
  ☐ identify the major organs of both the male and female reproductive systems in a dissected mammalian specimen
☐ Prepare your materials:
  ☐ preserved cat or fetal pig
  ☐ dissection tools and trays
  ☐ protective gear: gloves, safety eyewear, and apron
  ☐ storage container (if specimen is to be reused)
☐ Read the directions and safety tips for this exercise *carefully* before starting any procedure.

> **! safety first**
> Observe the usual precautions when working with a preserved specimen. Heed the safety advice accompanying preservatives used with your specimen. Use protective gloves while handling your specimen. Avoid injury with dissection tools. Use safety goggles to avoid injury during dissections. Dispose of or store your specimen as instructed. •

## *Study* TIPS

Obtain a labeled diagram or illustration of the male and female reproductive systems. Use these as you complete this lab. You may also want to use your cell phone camera to take pictures of your specimens and use them later for a study resource.

## A. REPRODUCTIVE ANATOMY OF THE MALE CAT

This activity challenges you to identify the major reproductive structures of the male cat. Assuming you have already removed the cat's skin (Lab Exercise 10—The Skin) and opened the abdominopelvic cavity (Lab Exercise 40—Dissection: Cardiovascular and Lymphatic System), you simply have to move the viscera out of the way (without cutting or removing them) to see many of the structures indicated in this activity. You may have to cut some additional skin or muscle tissue to see a few of the structures more clearly.

> **Hint** ▶ **Figure 52-1** illustrates many of the structures of the male cat listed in this activity. •

☐ 1 Locate the **scrotum,** a sac of skin between the hind limbs of the animal. Cut the scrotum open with your scissors or scalpel, taking care not to cut the structures inside the scrotum.

☐ 2 Identify the **testes,** the male *gonads* within the scrotum. Remove the connective tissue covering each testis to reveal the coiled **epididymis.** Note that the **vas (ductus) deferens** leads from the epididymis, through the *inguinal canal,* and into the abdominopelvic cavity.

☐ 3 Trace the course of one of the vas deferens into the abdominopelvic cavity to where it meets its partner from the other side. Note that they join together where they meet the **urethra.** At this location, a mass of glandular tissue called the **prostate gland** can be found.

☐ 4 Follow the course of the urethra toward the base of the **penis,** where a pair of small **bulbourethral glands** can be found.

☐ 5 Identify the **glans penis** at the distal tip of the penis.

# B. REPRODUCTIVE ANATOMY OF THE FEMALE CAT

This activity challenges you to identify the major reproductive structures of the female cat. Assuming you have already removed the cat's skin (Lab Exercise 10—The Skin) and opened the abdominopelvic cavity (Lab Exercise 40—Dissection: Cardiovascular and Lymphatic System), you simply have to move the viscera out of the way (without cutting or removing them) to see many of the structures indicated in this activity. You may have to cut some additional skin or muscle tissue to see a few of the structures more clearly.

> **Hint** ▸ **Figure 52-2** illustrates many of the structures of the female cat listed in this activity. •

☐ 1  Locate the tiny, lightly colored **ovaries,** the female *gonads,* located just posterior to the kidneys.

☐ 2  Identify the **oviduct,** with its distal *fimbriae,* that leads from each ovary toward a **uterine horn.** Each uterine horn is a lateral extension of the superior portion of the uterus.

☐ 3  Locate the main portion of the **uterus** along the midline of the body. The female reproductive tract continues posteriorly from the uterus as the **vagina** toward the outside of the body. The vagina ends at the *urogenital sinus.*

# C. REPRODUCTIVE ANATOMY OF THE MALE FETAL PIG

This activity challenges you to identify the major reproductive structures of the male fetal pig. Assuming you have already removed the fetal pig's skin (Lab Exercise 10—The Skin) and opened the abdominopelvic cavity (Lab Exercise 40—Dissection: Cardiovascular and Lymphatic System), you simply have to move the viscera out of the way (without cutting or removing them) to see many of the structures indicated in this activity. You may have to cut some additional skin or muscle tissue of the scrotum to see a few of the structures more clearly.

> **Hint** ▸ **Figure 52-3** illustrates many of the structures of the male fetal pig listed in this activity. •

☐ 1  Locate the **scrotum,** a sac of skin between the hind limbs of the animal. Cut the scrotum open with your scissors or scalpel, taking care not to cut the structures inside the scrotum.

☐ 2  Identify the immature **testes,** the male *gonads* within the scrotum. They should have descended into the scrotum late in fetal development. Remove the connective tissue covering each testis to reveal the coiled **epididymis.** Note that the **vas (ductus) deferens** leads from the epididymis, through the *inguinal canal,* into the abdominopelvic cavity.

☐ 3  Trace the course of one of the vas deferens into the abdominopelvic cavity to where it meets its partner from the other side. Note that they join together where they meet the **urethra.** A small, glandular **seminal vesicle** may be observed at the end of each vas deferens. Also at this location is a single mass of glandular tissue called the **prostate gland.**

☐ 4  Follow the course of the urethra toward the base of the **penis,** where a pair of small, elongated **bulbourethral glands** can be located.

# D. REPRODUCTIVE ANATOMY OF THE FEMALE FETAL PIG

This activity challenges you to identify the major reproductive structures of the female fetal pig. Assuming you have already removed the fetal pig's skin (Lab Exercise 10—The Skin) and opened the abdominopelvic cavity (Lab Exercise 40—Dissection: Cardiovascular and Lymphatic System), you simply have to move the viscera out of the way (without cutting or removing them) to see many of the structures indicated in this activity. You may have to cut some additional skin or muscle tissue to see a few of the structures more clearly.

> **Hint** ▸ **Figure 52-4** illustrates many of the structures of the female fetal pig listed in this activity. •

☐ 1  Locate the tiny, lightly colored **ovaries,** the female *gonads,* located just posterior to the kidneys.

☐ 2  Identify the **oviduct,** with its distal *fimbriae,* that leads from each ovary toward a **uterine horn.** Each uterine horn is a lateral extension of the superior portion of the uterus.

☐ 3  Locate the main portion of the **uterus** along the midline of the body. The female reproductive tract continues posteriorly from the uterus as the **vagina** toward the outside of the body. The vagina ends at the *urogenital sinus.*

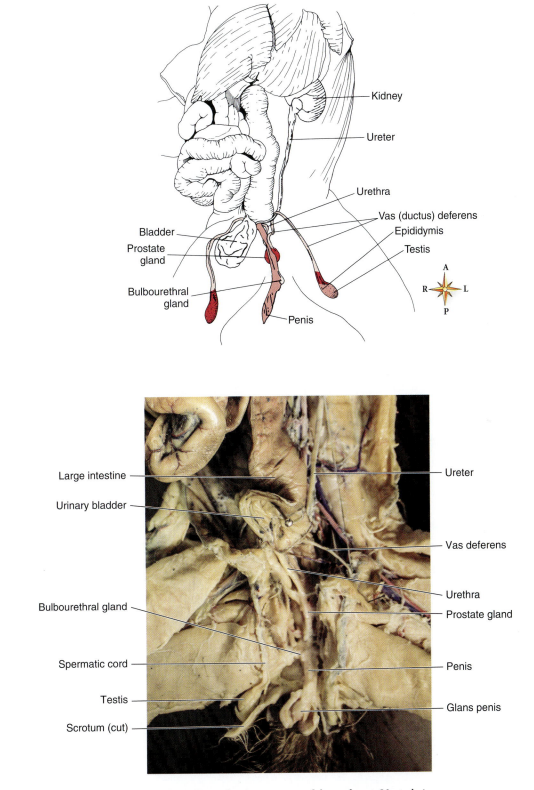

**FIGURE 52-1** Reproductive anatomy of the male cat. Ventral view.

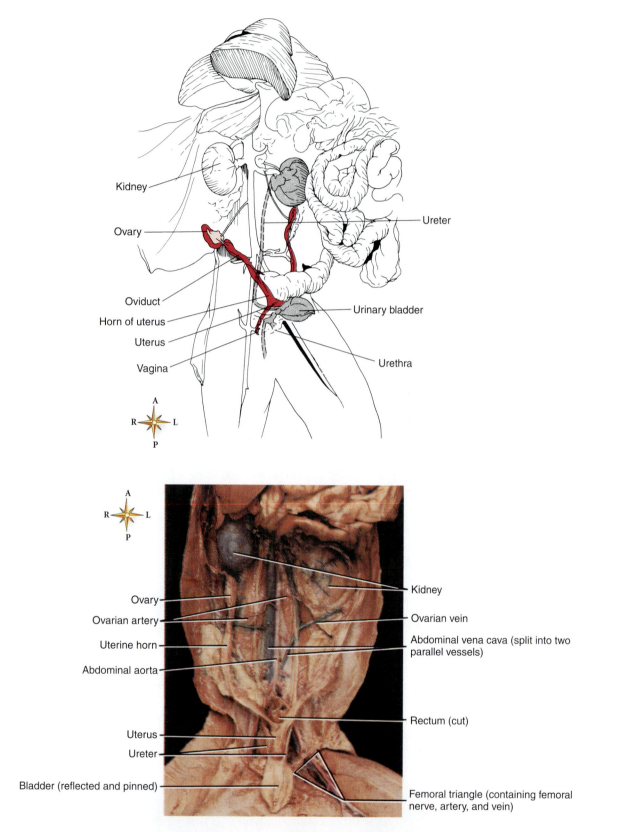

**FIGURE 52-2** Reproductive anatomy of the female cat. Ventral view.

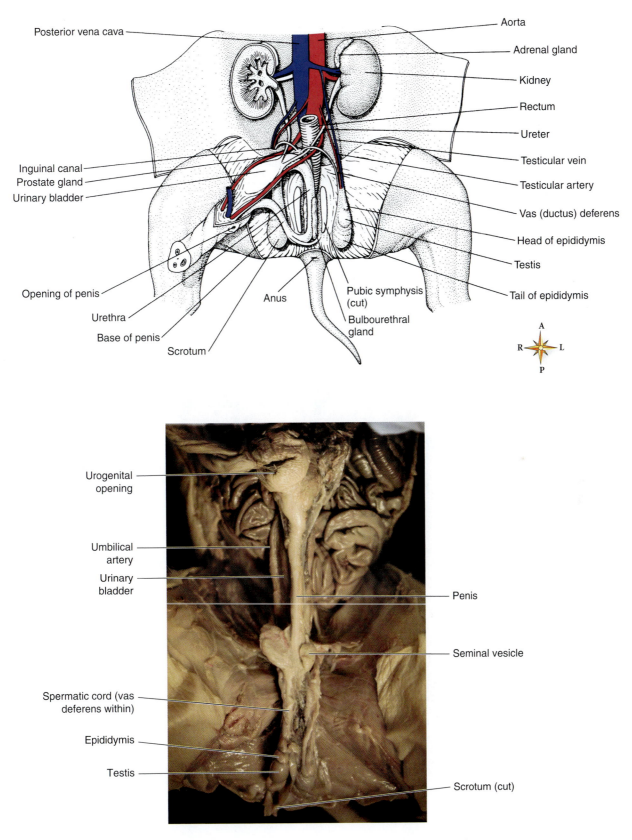

**FIGURE 52-3** Reproductive anatomy of the male fetal pig. Ventral view.

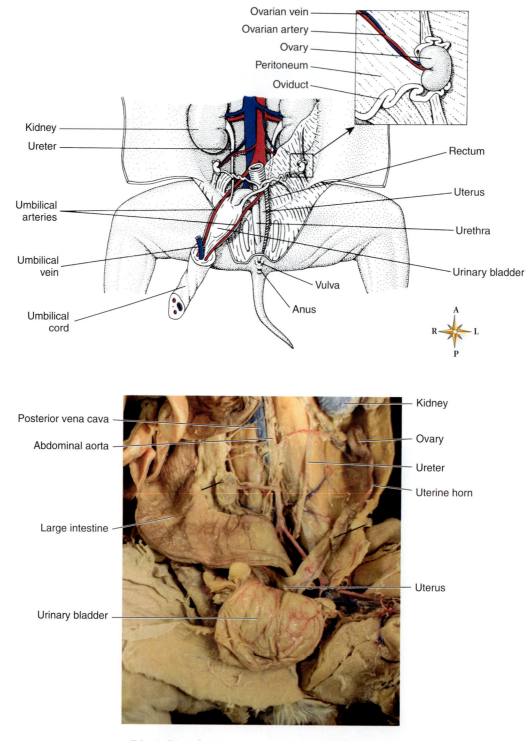

Ovarian vein

Ovarian artery

Ovary

Peritoneum

Oviduct

Kidney

Ureter

Rectum

Uterus

Umbilical arteries

Urethra

Umbilical vein

Urinary bladder

Vulva

Umbilical cord

Anus

A
R · L
P

Posterior vena cava

Kidney

Abdominal aorta

Ovary

Ureter

Uterine horn

Large intestine

Urinary bladder

Uterus

**FIGURE 52-4** Reproductive anatomy of the female fetal pig. Ventral view.

Name: _____ Date: _____ Section: _____

**LAB REPORT 52**
# Dissection: Reproductive Systems

## A. Male Cat Dissection Checklist

☐ scrotum
☐ testis
☐ epididymis
☐ vas (ductus) deferens
☐ urethra
☐ prostate gland
☐ penis
☐ bulbourethral glands
☐ glans penis

## B. Female Cat Dissection Checklist

☐ ovaries
☐ oviduct
☐ uterine horn
☐ uterus
☐ vagina

## C. Male Fetal Pig Dissection Checklist

☐ scrotum
☐ testis
☐ epididymis
☐ vas (ductus) deferens
☐ urethra
☐ seminal vesicle
☐ prostate gland
☐ penis
☐ bulbourethral glands

## D. Female Fetal Pig Dissection Checklist

☐ ovaries
☐ oviduct
☐ uterine horn
☐ uterus
☐ vagina

# Development

Human development begins from the moment of fertilization in the female reproductive tract. A sperm cell from the male parent and an oocyte from the female parent unite to form the first cell of the offspring. This cell divides (using mitosis) again and again, forming a mass of cells that will eventually become a mature human. This exercise presents a brief exploration of the stages of human prenatal development.

## BEFORE YOU BEGIN

☐ Read the appropriate chapter in your textbook.
☐ Set your learning goals. When you finish this exercise, you should be able to:
   ☐ describe the early stages of prenatal development
   ☐ identify major features of successive developmental stages in models or charts
☐ Prepare your materials:
   ☐ models or charts of early human development
   ☐ colored pens or pencils
☐ Read the directions for this exercise *carefully* before starting any procedure.

---

### *Study* TIPS

Develop a concept map that illustrates neonatal development. Include the major events that occur during each developmental period. For reference, use your textbook, lab manual, or the *A&P Survival Guide*.

---

## A. EARLY HUMAN DEVELOPMENT

Obtain a set of models or charts that depict the major stages of early human development. Use the guidance given here and in your textbook to identify the major events of early development and the landmark structures associated with each stage.

☐ 1 The original cell of an offspring is termed the **zygote.** About 18 to 36 hours after being formed by the fusion of an egg and sperm, this cell divides. The two daughter cells divide, then their daughter cells divide, and so on, until a ball of cells is formed. Locate a zygote in **Figure 53-1** and in your model.

☐ 2 When the ball has about 32 cells and a fluid-filled cavity called the **blastocele** forms, the offspring becomes a **blastocyst** and continues to grow. Most of the blastocyst is a single layer of cells called the **trophoblast,** but one portion also has an **inner cell mass** several cells thick. The inner cell mass develops into the embryo. The trophoblast develops into the **chorion,** which later becomes the placenta and the membranes that surround the embryo. Locate this stage in **Figure 53-1** and in your model.

☐ 3 Around 7 days after fertilization, the blastocyst begins **implantation** by digesting its way into the endometrium of the mother's uterus. During this stage the chorion (formerly the trophoblast) develops projections (**chorionic villi**) into the maternal blood supply. This close association of the embryonic and maternal tissue is called the **placenta.** Later in development the placenta allows diffusion of substances between the embryo's blood and the mother's blood. Identify this stage in **Figure 53-1**, then in your model.

☐ 4 By day 11, several features have become visible. Identify these features in **Figure 53-1**, then in a model or chart:
   ☐ **Connecting stalk**—This narrow piece of tissue connects the inner cell mass to the developing placenta (it will eventually develop into the **umbilical cord**)
   ☐ **Amniotic cavity**—A fluid-filled space within the cell mass, it lies within a layer of cells called the **ectoderm** (this cavity will eventually develop into the **amniotic sac** surrounding the embryo)
   ☐ **Yolk sac**—Another fluid-filled space within the cell mass, this one lies within a layer of cells called the **endoderm**
   ☐ **Embryonic disk**—This flat sheet of tissue is formed by adjacent layers of the endoderm and ectoderm

☐ 5 By day 14, more features of the developing offspring become apparent, as seen in **Figure 53-1**:
   ☐ **Primitive streak**—This is a thickened line in the endoderm at the center of the elongated embryonic disk (the embryo will eventually form around the streak; the **notochord** extends from the cephalic end of the streak)
   ☐ **Mesoderm**—This new *germ layer* arises between the ectoderm and the endoderm

☐ 6 Follow the developmental stages represented in your model or chart, noting that the three germ layers (endoderm, mesoderm, and ectoderm) develop into different tissues and organs. Using your textbook or a reference book, determine which major tissues or organs are derived from each of the three germ layers. Report your findings in the Lab Report at the end of this exercise.

☐ 7 After about 60 days, the cartilage and membranous tissue of the embryonic skeleton begin to ossify. From this point onward, the offspring is termed a **fetus.** Examine a model or chart of a full-term fetus. Because its organs and systems have developed sufficiently, the full-term fetus is able to survive outside the uterus. Locate these structures associated with the fetus:

☐ **Amniotic sac**—A fluid-filled space derived from the amniotic cavity, it cushions the developing offspring. At **parturition**, the sac breaks and spills its contents through the birth canal (vagina).

☐ **Placenta**—The organ originally formed by the chorionic villi's implantation into the endometrial lining of the uterus, the mature placenta allows an exchange of materials between the fetal circulatory system and the maternal circulatory system. Examine the anatomical relationship of these two sets of vessels.

☐ **Umbilical cord**—Derived from the connecting stalk observed in earlier stages, the umbilical cord includes **umbilical arteries** and an **umbilical vein**. Lab Exercise 38—The Circulatory Pathway describes how these vessels fit into the fetal circulatory plan. The umbilical cord serves as a connection between the developing offspring and the mother's body.

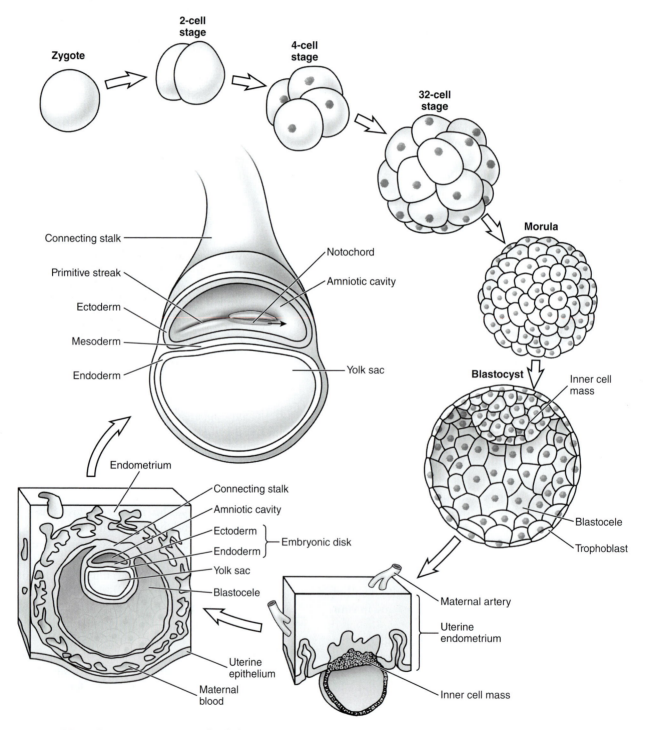

**FIGURE 53-1**   Representative stages of early human development. This figure may be colored in with colored pens or pencils.

# *clinical application: ultrasonography*

**Ultrasonography** is a method of imaging body parts by using high-frequency sound (ultrasound) waves. Usually a handheld wand that emits ultrasonic waves is placed against the patient's skin. The waves pass through different tissues at different speeds, and some waves are reflected back to the wand. Sensors in the wand detect the reflected ultrasound and relay the information to a computer capable of constructing an image based on the reflections. The image produced by this technique is called a *sonogram*.

An advantage of this technique is that ultrasound energy is not as dangerous as the radiation used in computed tomography (CT) scans and regular radiographs. This is one reason that it has become popular among obstetric physicians. Developing fetal tissue that could be harmed by x-rays is not harmed by ultrasound waves; therefore the development of a particular embryo or fetus can be monitored by ultrasonography **(Figure 53-2)**.

Sonograms can be used to determine whether there is a single fetus or multiple fetuses. Abnormally developed structures are sometimes visible. Late in development, male or female genitals can sometimes be seen.

Occasionally, measurements of a fetus at various stages are made. The normal rate of development of organs can then be verified. Usually the age of a fetus is given as the clinical age rather than the developmental age used earlier in this exercise. The clinical age is the time since the mother's last menstrual period, whereas the developmental age is the time since fertilization.

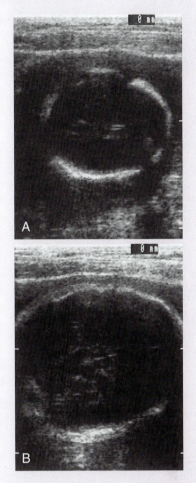

**FIGURE 53-2** Sonogram **A** shows the outline of a fetal skull at 153 days (clinical age). Compare it with sonogram **B**, the skull of the same fetus shown at 206 days.

1. The size of the skull has increased by what percentage during the 53 days between sonograms?

2. Which structures in particular do you think are likely to show up well in a sonogram?

3. What other applications for ultrasonography can you think of?

4. What advantages do CT scans, magnetic resonance (MR) images, and regular radiographs have over sonograms?

Name: _____ Date: _____ Section: _____

# LAB REPORT 53
# Development

## Identify

1. _____
2. _____
3. _____
4. _____
5. _____
6. _____
7. _____
8. _____
9. _____
10. _____
11. _____
12. _____
13. _____
14. _____
15. _____
16. _____
17. _____
18. _____
19. _____
20. _____
21. _____
22. _____
23. _____
24. _____
25. _____

**Identify** (see Figure 8-1 in the textbook to assist you with identifying the germ layer from which each of these organs or tissues is derived; write "endo" for endoderm, "ecto" for ectoderm," and "meso" for mesoderm)

1. Adrenal medulla
2. Anterior pituitary gland
3. Bones of the face
4. Bones (except those of the face)
5. Brain
6. Cardiovascular organs
7. Dermis
8. Epidermis
9. Gonads
10. Kidney ducts and bladder
11. Kidneys
12. Lens and cornea of the eye
13. Lining of the gastrointestinal tract
14. Linings of the hepatic and pancreatic ducts
15. Lining of the lungs
16. Muscle (except skeletal muscles of the head)
17. Nasal cavity
18. Outer ear
19. Parathyroid gland
20. Skeletal muscles of the head
21. Spinal cord
22. Thymus gland
23. Thyroid gland
24. Tonsils
25. Tooth enamel

## Multiple Choice

1. _____
2. _____
3. _____
4. _____
5. _____
6. _____
7. _____

## Multiple Choice (choose the best response)

1. The cell formed by the union of a sperm cell and an oocyte is called a
   a. blastula
   b. morula
   c. zygote
   d. blastocele
   e. b and c are correct
2. The connecting stalk seen early in human development later becomes the
   a. placenta
   b. amniotic sac
   c. umbilical cord
   d. yolk sac
3. The earliest developmental stage at which one can see a fluid-filled extracellular cavity is
   a. morula
   b. blastula
   c. zygote
   d. fetus
   e. 4-cell stage
4. The organ through which maternal and fetal blood exchange materials is the
   a. umbilical cord
   b. umbilical vein
   c. amniotic sac
   d. amniotic cavity
   e. placenta
5. The process by which a baby is born is called
   a. gestation
   b. pregnancy
   c. differentiation
   d. parturition
   e. lactation
6. The chorion is derived from the
   a. trophoblast
   b. blastocele
   c. inner cell mass
7. The embryo develops around a thickened line in the endoderm called the
   a. trophoblast
   b. inner cell mass
   c. notochord
   d. primitive streak

# 54
# Genetics and Heredity

Heredity, or the inheritance of traits, falls into the realm of **genetics.** Genetic information in the form of **genes** is found in the 23 DNA molecules that you inherit from your mother and the 23 that you inherit from your father. This information is passed from generation to generation. Twenty-two pairs of your DNA molecules, or *chromosomes*, are called **autosomes.** The remaining pair is called the **sex chromosomes.** One member of each pair is inherited from one parent, and the other member of the pair from the other parent.

Each autosome in a pair is homologous to the other member of the pair. With this arrangement, a person has two genes for every inherited characteristic. If one gene is always expressed, whether or not its mate is the same gene, geneticists call that gene **dominant.** A gene that is not expressed when its mate is different is termed a **recessive gene.**

All the genes in all the chromosomes taken together are called the **genome.** The study of all the genes in the human species, the *human genome*, is called **genomics.**

## BEFORE YOU BEGIN

☐ Read the appropriate chapter in your textbook.
☐ Set your learning goals. When you finish this exercise, you should be able to:
　☐ identify the phenotype and possible genotypes of selected human characteristics
　☐ distinguish between the concepts of *dominant* and *recessive* as applied to inherited traits
　☐ predict probabilities of phenotypes and genotypes in offspring given the parents' genotypes or phenotypes
☐ Prepare your materials:
　☐ taste papers: control, sodium benzoate, phenylthiocarbamide (PTC), thiourea
　☐ BIOHAZARD container
　☐ blood typing materials (if not done previously in Lab Exercise 34—Blood)
☐ Read the directions and safety tips for this exercise *carefully* before starting any procedure.

### *Study* TIPS

Review the principles of inheritance. Understand how a Punnett square is used to determine the probabilities of producing offspring with specific gene combinations. For reference, use your textbook or the *A&P Survival Guide*.

## A. HUMAN PHENOTYPES AND GENOTYPES

For any particular inherited characteristic, a **phenotype** can be identified. The phenotype is the characteristic actually expressed in an individual. A person's **genotype** is a statement of both genes that influence a particular trait. For example, the phenotype *albinism* is a recessive genetic condition in which a person lacks skin pigmentation. By convention, a gene for normal skin pigmentation, which is dominant, is represented as A. The recessive gene is represented as *a*. A person with normal skin pigmentation may have the genotype AA, meaning that both genes are normal. Geneticists may state that the person's genotype is **homozygous,** meaning that both genes are the same. A person with normal skin color could also have the Aa genotype. Because A is dominant, the abnormal *a* is not expressed. This person has a **heterozygous** genotype, meaning that the two genes are not the same. A person with albinism must have the genotype *aa* because it is the only genotype that allows the recessive abnormal trait to be expressed. Thus the normal phenotype (normal skin color) may be associated with the genotype *AA* or the genotype *Aa*. The abnormal phenotype (albinism) is always associated with the genotype *aa*.

### *Study* TIPS

Skin color is determined by a number of factors, both environmental and genetic. A number of different *A* genes exist in the human gene pool, and genes at other locations in the DNA may influence skin color. The example of albinism is simple when first observed, but the effect of multiple factors complicates things. This is typical of inherited characteristics. Often, many genes influence one particular trait. Also, many environmental factors can affect inherited characteristics.

### ⓘ *safety first*

The first step of this activity involves tasting test papers. Be sure to use clean procedures in handling the papers before and after you taste them. Used papers are to be placed in a BIOHAZARD container immediately after use. •

☐ 1 The first set of characteristics to be determined involves the sense of taste. It is known that the ability to taste certain compounds depends on the presence of certain genes. You will taste papers that have each been impregnated with a different compound. Before starting, put a piece of CONTROL test paper on your tongue and chew it. If you taste something, you must be sensitive to a compound in the paper itself. Ignore that particular taste in the tests that remain or try to sense *differences* in taste.

Start here:

☐ **Sodium benzoate test**—Ability to taste something sweet, salty, or bitter in the paper is dominant

☐ **PTC test**—Ability to sense a bitter taste is dominant

☐ **Thiourea test**—Ability to taste something bitter is dominant

☐ 2 The next set of determinations involves anatomical characteristics of your hand:

☐ **Bent little finger (Figure 54-1, A)**—Place your relaxed hand flat on the lab table. If the distal phalanx of the little (fifth) finger bends toward the fourth finger, you have the dominant trait.

☐ **Middigital hair** (see **Figure 54-1, B**)—Dorsal hair on the skin over the middle phalanges of the hand is dominant.

☐ **Hitchhiker's thumb** (see **Figure 54-1, D**)—If you can hyperextend the distal joint of the thumb noticeably, you have the recessive trait.

☐ 3 Determine your phenotype for these facial features:

☐ **Pigmented anterior of the iris**—If you have pigment on the anterior *and* posterior of the iris, your eyes are green, brown, black, or hazel. If you lack pigment on the anterior aspect of the iris, your eyes are blue or gray. Anterior pigmentation is dominant.

☐ **Attached earlobes (Figure 54-2, A)**—If the inferior, fatty lobe of the ear is attached rather than free, you have the recessive trait.

☐ **Widow's peak** (see **Figure 54-2, B**)—Assuming you have a hairline, if it is straight across the forehead, you have the recessive trait. If it forms a downward point near the midline, you have the dominant "widow's peak."

☐ **Tongue roll** (see **Figure 54-2, C**)—Try to curl your tongue as you extend it from your mouth. If you can't curl it, you have the recessive trait.

☐ **Freckles**—If your face has a scattering of freckles, you have the dominant form of this characteristic. If your face is free of freckles, you have the recessive condition.

☐ 4 There are two dominant genes for ABO blood types. One is $I^A$, which signifies the presence of the A antigen. The other is $I^B$, signifying the B antigen. The recessive gene is $i$, signifying neither ABO antigen. A person with the phenotype TYPE A has the genotype $I^AI^A$ or $I^Ai$. A person with the phenotype TYPE B has the genotype $I^BI^B$ or $I^Bi$. A person with phenotype TYPE AB has the genotype $I^AI^B$. Phenotype TYPE O requires the genotype $ii$. Record your phenotype (your ABO blood type). What is your genotype?

**Hint** ▶ If you have not already typed your blood (as instructed in Lab Exercise 34—Blood), consult your health records or perform the typing now. Refer to Lab Exercise 34 and heed the safety advice given there. ●

☐ 5 The Rh blood type is also determined by genetics. Presence of the Rh antigen is dominant. What is your Rh blood type? What is your genotype?

## *Study* TIPS

This section instructs you to determine your phenotype for a variety of easily observed characteristics. Once you know your phenotype, you can determine your possible genotypes. For example, if you are an albino, you know that your genotype is *aa*. If you do not have albinism, your genotype is either *AA* or *Aa*. For each characteristic described, record your phenotype and possible genotypes in the Lab Report at the end of this exercise.

## B. PROBABILITIES OF INHERITANCE

Now that you have a grasp of the concept of phenotype and genotype, you can move on to the concept of **probability.** Probability is the likelihood of a certain outcome in a particular event. As applied to human genetics, probability refers to the likelihood that the offspring of a particular set of parents will have a certain inherited condition. **Genetic counselors** work with prospective parents to determine their possible genotypes for a variety of traits. Then they predict the probability of their children having those traits. In this way, parents can anticipate possible abnormal genetic conditions.

In this exercise, we will use a simple method developed by the English geneticist Reginald Punnett. He devised a simple grid, or **Punnett square,** with which one can easily predict simple ratios of genetic probability.

First, an example:

*The father has freckles; the mother does not. The gene for freckles is* **F,** *and the gene for no freckles is* **f.** *The father's genotype is either* **FF** *or* **Ff.** *The mother's genotype must be* **ff.** *We must consider two different outcomes in offspring from this couple because the father has two possible genotypes for this trait. First, let's assume that the father's genotype is* **FF.** *Any sperm cell contributed by the father will have the* **F** *gene. Any oocyte contributed by the mother will have the* **f** *gene. We place the possible gene for the father in the left margin of the Punnett square and the mother's in the top margin:*

*To use the Punnett square, start in the top left square of the grid. Combine the gene from the left margin with the gene from the top margin:*

We now have one possible offspring genotype: **Ff.** We then do the same for the other three squares of the grid:

Our Punnett square predicts a 100% probability that any one off-spring will have genotype **Ff** and therefore have the freckled phenotype. The second possibility is that the father's genotype is **Ff.** We can set up a Punnett square for this possibility:

We can now fill in the grid with possible offspring genotypes, combining the gene from the left margin with the gene from the top margin for each square:

```
      f    f
  F | Ff | Ff |
  f | ff | ff |
```

Because half the squares have genotype **Ff,** there is a 50% probability that any offspring of this couple will have the **Ff** genotype (and therefore the freckled phenotype). There is also a 50% probability that the genotype will be **ff,** expressed as the nonfreckled phenotype. If the phenotypes or genotypes of the client's parents and grandparents are known, a genetic counselor might construct a **pedigree,** or family tree, to determine the father's likely genotype. If that information is unavailable, the parents will have to deal with the two different probabilities.

Now that you have seen an example, try your hand at the following problems. Draw your Punnett squares and record your results in the Lab Report at the end of this exercise.

☐ 1 Huntington disease (Huntington chorea) is a degenerative nerve disorder with a genetic basis that becomes apparent after about 40 years of age. The abnormal gene that produces this disease, *H*, is dominant. The normal, recessive gene is *h*. One of Heather's parents has Huntington disease, and the other does not. Can you predict the highest probability that Heather will develop Huntington disease later in her life?

☐ 2 Kevin has Rh-positive blood. Christine has Rh-negative blood. Their first child, Andrew, has Rh-positive blood. Both of Kevin's parents have Rh-positive blood. What is the probability that the second child Kevin and Christine are expecting will be Rh negative?

☐ 3 Leo's father has albinism, but Leo does not. Cleo's father has albinism, but she does not. If Leo and Cleo have a child, what is the probability that it will have albinism? What is the probability that their second child will have albinism? Their third child?

☐ 4 In the ABO blood typing system, Mario is type O, and Ana is type AB. What ABO blood types might their children have?

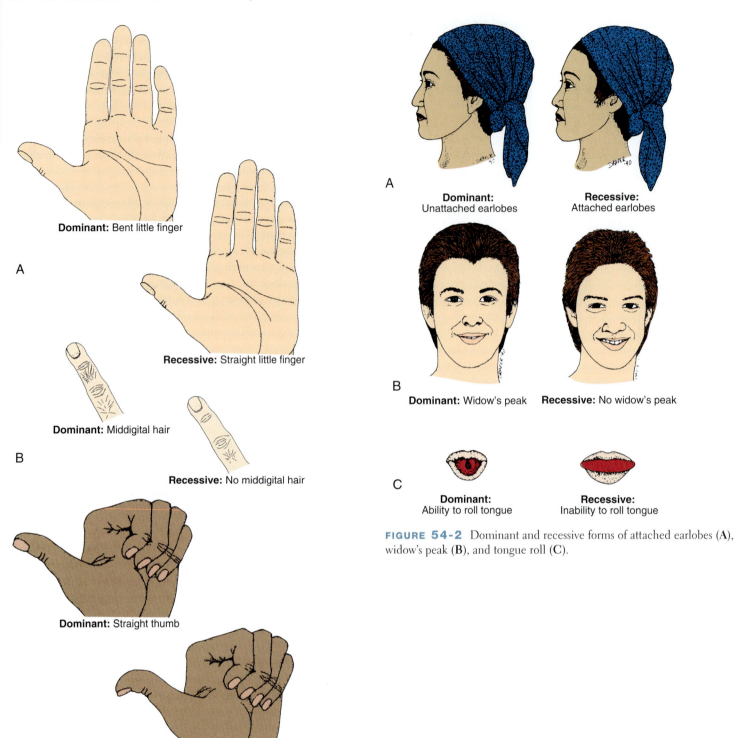

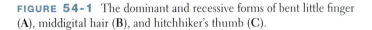

**FIGURE 54-1** The dominant and recessive forms of bent little finger (**A**), middigital hair (**B**), and hitchhiker's thumb (**C**).

**FIGURE 54-2** Dominant and recessive forms of attached earlobes (**A**), widow's peak (**B**), and tongue roll (**C**).

Name: _____ Date: _____ Section: _____

## LAB REPORT 54
# Genetics and Heredity

| TRAIT | DOMINANT GENE(S) | RECESSIVE GENE | YOUR PHENOTYPE | POSSIBLE GENOTYPES |
|---|---|---|---|---|
| Sodium benzoate taste | S | s | | |
| PTC taste | P | p | | |
| Thiourea taste | T | t | | |
| Bent little finger | L | l | | |
| Middigital hair | M | m | | |
| Hitchhiker's thumb | H | h | | |
| Pigmented anterior of iris | I | i | | |
| Attached earlobes | A | a | | |
| Widow's peak | W | w | | |
| Tongue roll | R | r | | |
| Freckles | F | f | | |
| ABO blood type | IA, IB | i | | |
| Rh blood type | D | d | | |

*PTC*, Phenylthiocarbamide.

1. What is the highest probability that Heather will develop Huntington disease? ___%

2. What is the probability that Kevin and Christine's second child will have Rh-negative blood? _____ or _____% (*Hint: There are two possibilities.*)

3. What is the probability that Leo and Cleo's first child will be albino? _____%
   Their second child? _____%
   Their third child? _____%

4. What are the possible ABO blood types of Mario and Ana's children? _____

# LAB EXERCISE 55
# The Whole Body

This exercise is intended as a possible summary or synthesis activity for the laboratory course. Whether your instructor will invite you to complete this exercise will depend on the course schedule and the availability of the required materials.

The first activity of this exercise suggests that you observe a live or videotaped demonstration of human anatomy using a human body. The second activity presents a special focus on transverse (horizontal) sectional anatomy, which is becoming increasingly important in practical applications of human anatomy.

## BEFORE YOU BEGIN

☐ Set your learning goals. When you finish this exercise, you should be able to:
  ☐ describe the overall body plan of the human
  ☐ identify the major organs of the human in a previously dissected cadaver or in a human body model or chart
  ☐ identify structures in selected transverse sections of the human body
☐ Prepare your materials:
  ☐ prosected human cadaver (for demonstration, if available) or human dissection videotape
  ☐ plastinated preparation (or chart): *human thorax, horizontal section (just above heart)*
  ☐ plastinated preparation (or chart): *human abdomen, horizontal section (pancreas level)*
  ☐ demonstration pointer
  ☐ hand lens
  ☐ colored pencils or pens
  ☐ protective gear: gloves, safety eyewear, and apron (optional)
☐ Read the directions and safety tips for this exercise *carefully* before starting any procedure.

### Study TIPS

The dissection of the human cadaver is a compilation of your study of human anatomy. The chance to learn from a cadaver is invaluable. You may want to obtain an atlas of the human body such as the *Brief Atlas* to use as you proceed through your dissection. This reference will assist you in your identification of the various structures of the cadaver.

## A. THE HUMAN CADAVER

Advanced laboratory courses in human anatomy use the human cadaver as the basic dissection specimen. A **cadaver** is an embalmed corpse. If available, view a live or videotaped demonstration using a prosected human cadaver. A prosected specimen is one that has been previously dissected and prepared for anatomical demonstrations.

### safety first

Do not touch the cadaver unless invited to do so by the demonstrator. If you are invited to touch it, use gloved hands. Cadavers are normally fixed in preservative, which is a toxic substance. Do not inhale the fumes that are immediately above the specimen. Excuse yourself from the demonstration if you feel ill or if you may be sensitive to preservatives such as formalin. Pregnant women should avoid contact with formalin fumes. Use safety goggles to avoid injury during dissections. •

## B. TRANVERSE SECTIONAL ANATOMY

Clinical and research institutions are seeing an increased use of computed tomography (CT) scans, magnetic resonance imaging (MRI), ultrasonography, and other advanced imaging techniques. Each of these techniques requires a basic knowledge of human anatomy as seen in a transverse, or horizontal, section.

Your studies so far have emphasized the sagittal and frontal aspects of anatomy. Now that you are familiar with the essentials of human anatomy from those perspectives, it is time to apply that knowledge to locating structures in transverse preparations.

### safety first

Plastinated specimens are natural tissues that have been embedded in plastic. Even though they long outlast normal embalmed specimens, they are very fragile. If you handle them at all, be careful not to damage them. Always use a demonstration pointer when pointing to structures, never a pen or pencil. Do not touch the specimen with your pointer; just point at it. •

☐ 1 Obtain a plastinated transverse section of a human torso (Figure 55-1) just superior to the heart (at about T4). Try to locate these structures:
  ☐ **Aorta**—If the aorta is present, what section is it (ascending, arch, or descending)?
  ☐ **Superior vena cava**—Can you identify this vein by its position and the thinness of its wall?
  ☐ **Esophagus**—This is a small (and perhaps flattened) muscular tube anterior to a vertebra.
  ☐ **Trachea**—Are any of the C-shaped cartilage rings visible in the tracheal wall?

☐ **Thymus**—It is posterior to the sternum.

☐ **Lungs**—Can you distinguish any portions of the respiratory tract within the lung tissue? Identify the parietal and visceral pleurae and the pleural cavity.

☐ **Sternum**—This structure is in the anterior wall of the thoracic cavity.

☐ **Ribs**

☐ **Thoracic vertebrae**—What features of the vertebrae are visible in your specimen?

☐ **Spinal cord**—It is inside the vertebral foramen. What features of the spinal cord can you identify?

☐ **Skeletal muscles**—They form the wall of the thoracic cavity.

☐ **Subcutaneous tissue**—Identify the areolar and/or adipose tissue under the skin.

☐ **Integument**—Can you distinguish between the dermis and epidermis? Are there any other organs or structures visible in your specimen? If the upper arms are present in your specimen, identify the humerus and other arm structures.

☐ 2 **Obtain a plastinated transverse section of the human abdomen (Figure 55-2) at the level of the pancreas (about T12). Try to locate these structures:**

☐ **Vertebra**—What features are visible in this preparation?

☐ **Spinal cord**—Can you distinguish white and gray matter?

☐ **Kidneys**—Are they the same size? Why or why not? Identify the renal fat pad around each kidney. Can you distinguish between the renal cortex and the renal medulla?

☐ **Aorta**—The abdominal aorta may be present just anterior to the body of the vertebra.

☐ **Inferior vena cava**—This large vein is normally next to the aorta.

☐ **Pancreas**—This glandular tissue is near the stomach.

☐ **Stomach**—This large muscular organ should be present in the left portion of the specimen. Can you identify the rugae (large folds) of the gastric mucosa?

☐ **Liver**—This is the dominant feature of your specimen, a large mass of tissue to the right and anterior of the abdominal cavity.

☐ **Colon**—Is any portion of the colon visible in your specimen? If so, which section is it?

☐ **Peritoneum**—Can you distinguish between the parietal and visceral portions of the peritoneum? Identify the peritoneal cavity. Are any mesenteries distinguishable in your specimen? The lesser omentum?

☐ **Ribs**

☐ **Abdominal muscles**—Which abdominal muscles could be present in your section?

☐ **Subcutaneous tissue**—Is the adipose tissue in this layer of the same thickness all the way around the abdominal wall?

☐ **Integument**—Are the dermis and epidermis distinguishable in this section?

---

**Hint** ▶ Each specimen is unique, just as every person's body is unique. Therefore some structures listed may not be present (or identifiable) in your specimen. Use a hand lens to examine the detail of the structures you have found. Refer to **Figures 55-3 through 55-6**, which show color photographs of transverse human cadavers. Also refer to **Figures 55-5** and **55-6**. These plates show transverse sections of a cadaver's abdomen. •

Use colored pens or pencils to shade in both the figure and the labels. Each red numeral in the figure corresponds to a matching red numeral following the appropriate label.

AORTA 1
SUPERIOR VENA CAVA 2
ESOPHAGUS 3
TRACHEA 4
THYMUS 5
LUNG 6
PLEURAL CAVITY 7

STERNUM 8
RIB 9
VERTEBRA 10
SPINAL CORD 11
SKELETAL MUSCLE 12
SUBCUTANEOUS TISSUE 13
INTEGUMENT 14

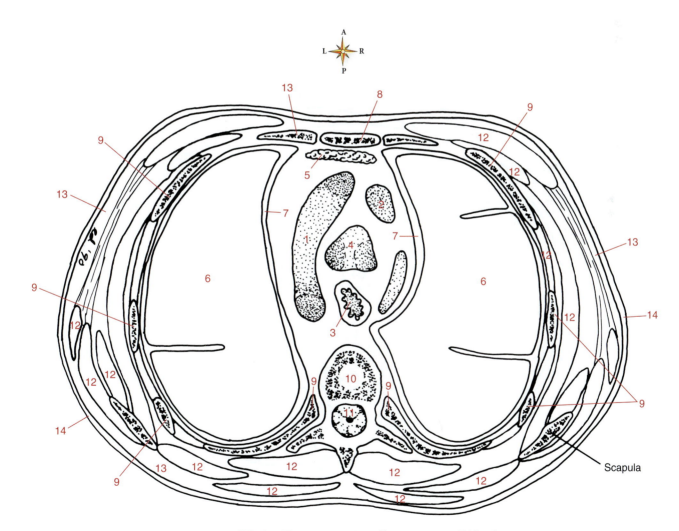

**FIGURE 55-1** Transverse section of human torso at T4 level.

## Coloring Exercises: Transverse Abdominal Section

Use colored pens or pencils to shade in both the figure and the labels. Each red numeral in the figure corresponds to a matching red numeral following the appropriate label.

VERTEBRA 1
SPINAL CORD 2
KIDNEY 3
AORTA 4
INFERIOR VENA CAVA 5
PANCREAS 6
STOMACH 7

LIVER 8
COLON 9
PERITONEAL CAVITY 10
RIB 11
ABDOMINAL MUSCLE 12
SUBCUTANEOUS TISSUE 13
INTEGUMENT 14

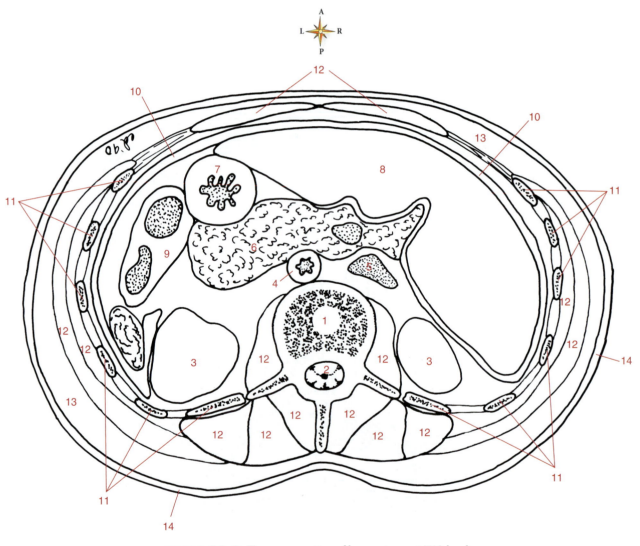

**FIGURE 55-2** Transverse section of human torso at T12 level.

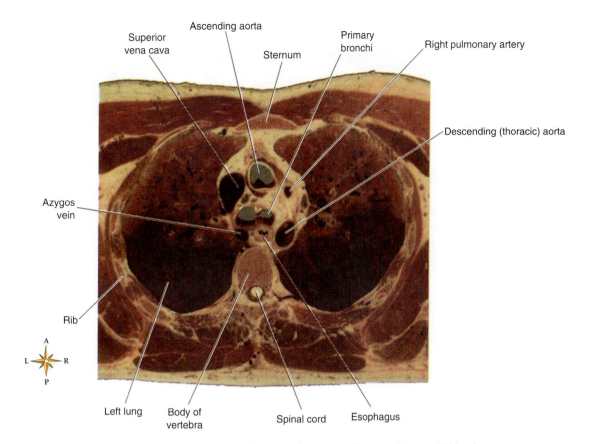

**FIGURE 55-3** Transverse (horizontal) section of human thorax at T4 level.

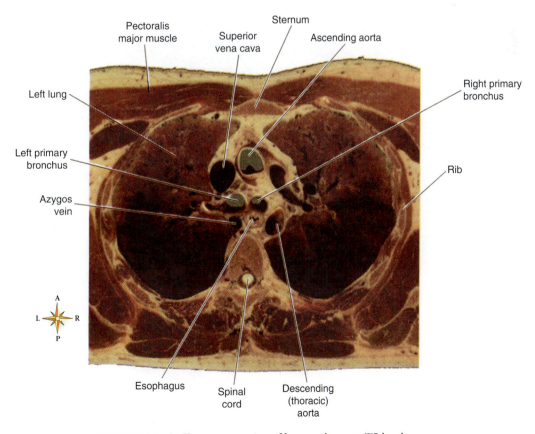

**FIGURE 55-4** Transverse section of human thorax at T5 level.

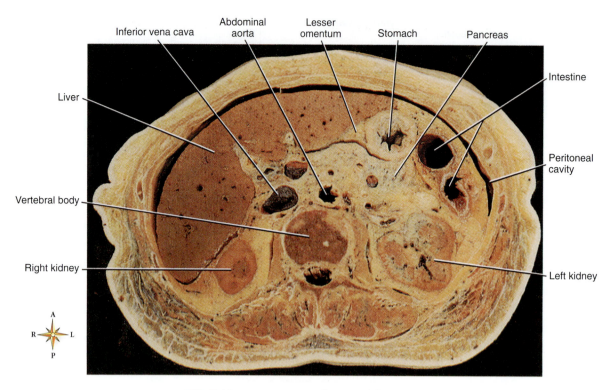

**FIGURE 55-5** Transverse section of human abdomen at T12 level.

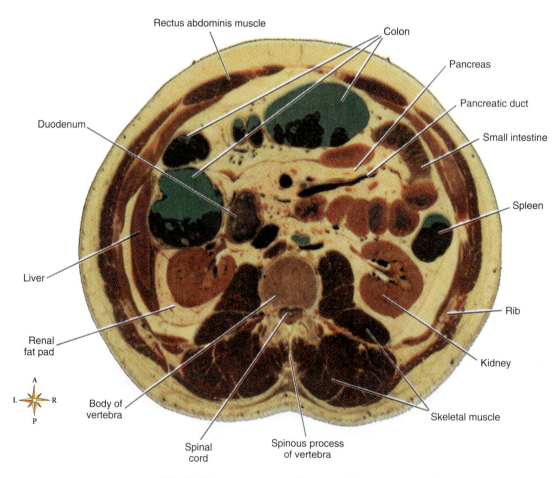

**FIGURE 55-6** Transverse section of human abdomen at L2 level.

Name: _____ Date: _____ Section: _____

**LAB REPORT 55**
# The Whole Body

Sketch and label your thorax (transverse section) specimen:

Sketch and label your abdomen (transverse section) specimen:

Name: _____ Date: _____ Section: _____

# Microscopic Observations

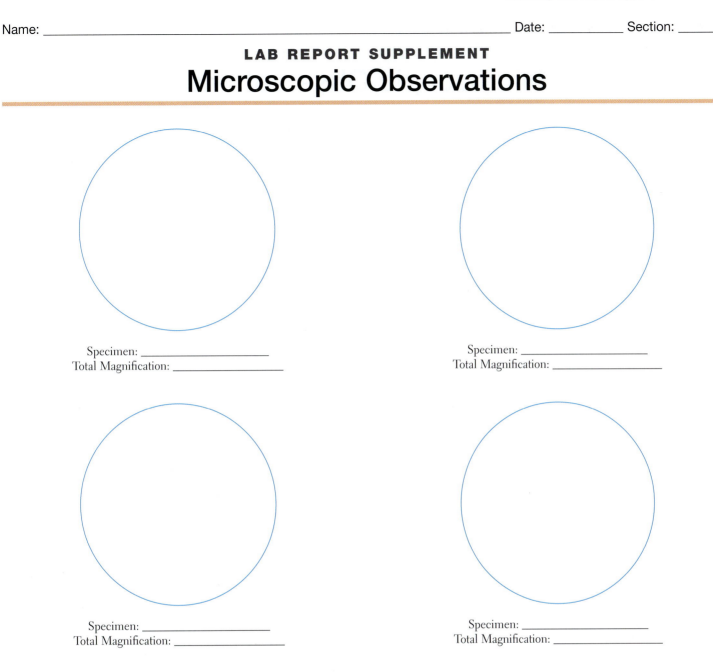

Specimen: _____
Total Magnification: _____

Specimen: _____
Total Magnification: _____

Specimen: _____
Total Magnification: _____

Specimen: _____
Total Magnification: _____

**Notes:**

Name: _____ Date: _____ Section: _____

# Microscopic Observations

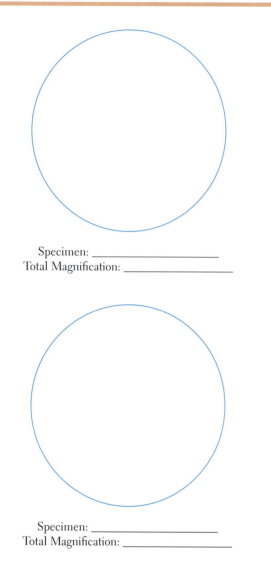

Specimen: _____
Total Magnification: _____

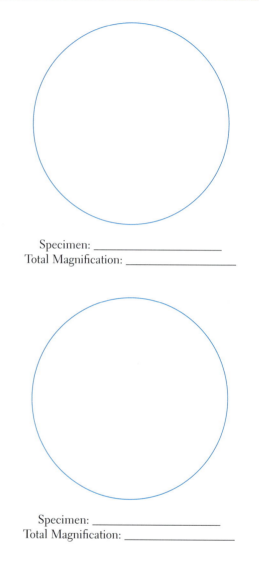

Specimen: _____
Total Magnification: _____

Specimen: _____
Total Magnification: _____

Specimen: _____
Total Magnification: _____

**Notes:**

Name: _____   Date: _____   Section: _____

**LAB REPORT SUPPLEMENT**
# Microscopic Observations

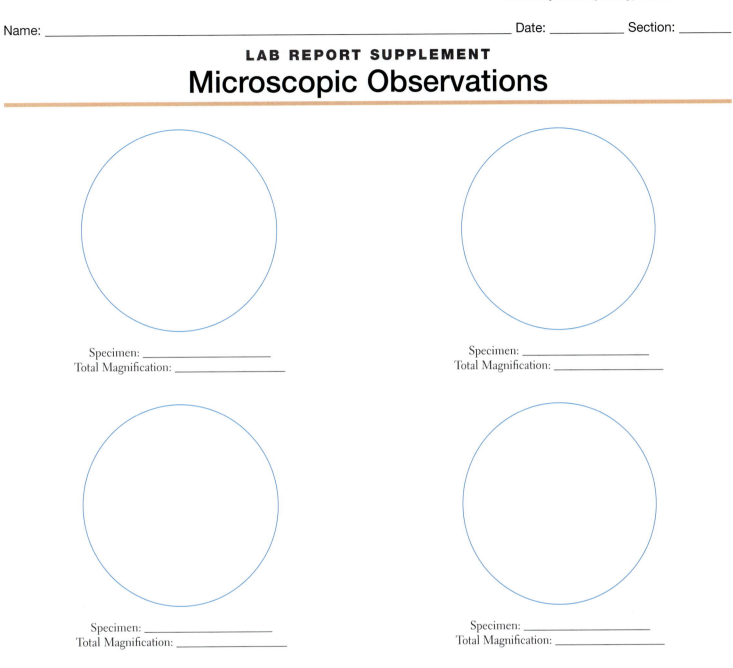

Specimen: _____
Total Magnification: _____

Specimen: _____
Total Magnification: _____

Specimen: _____
Total Magnification: _____

Specimen: _____
Total Magnification: _____

**Notes:**

Name: _____ Date: _____ Section: _____

**LAB REPORT SUPPLEMENT**
# Microscopic Observations

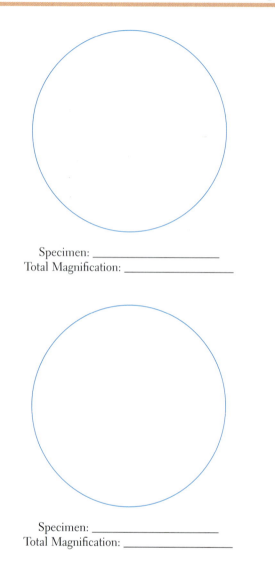

Specimen: _____
Total Magnification: _____

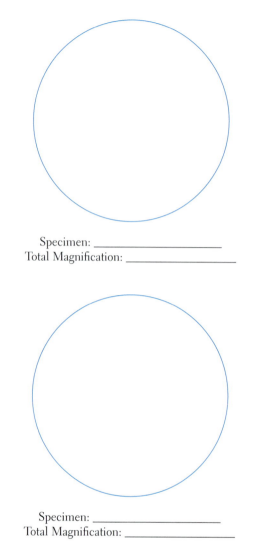

Specimen: _____
Total Magnification: _____

Specimen: _____
Total Magnification: _____

Specimen: _____
Total Magnification: _____

**Notes:**

Name: _____ Date: _____ Section: _____

# Microscopic Observations

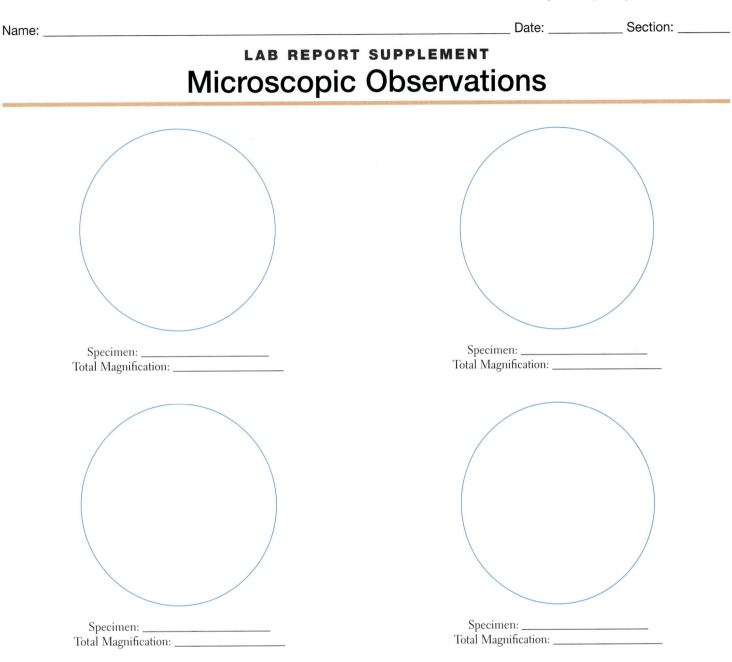

Specimen: _____
Total Magnification: _____

Specimen: _____
Total Magnification: _____

Specimen: _____
Total Magnification: _____

Specimen: _____
Total Magnification: _____

**Notes:**

Name: _____ Date: _____ Section: _____

# Microscopic Observations

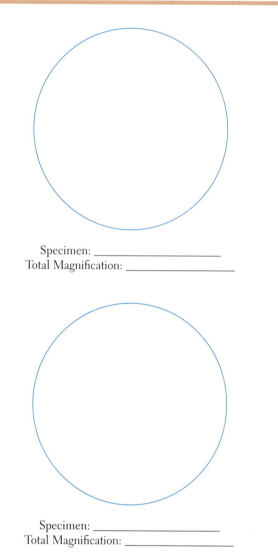

Specimen: _____
Total Magnification: _____

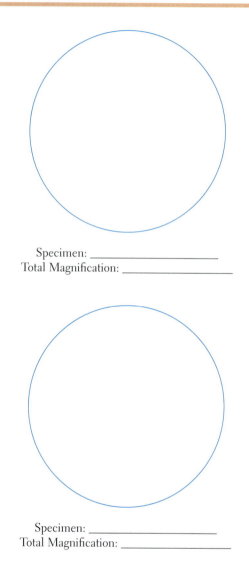

Specimen: _____
Total Magnification: _____

Specimen: _____
Total Magnification: _____

Specimen: _____
Total Magnification: _____

**Notes:**

Name: _____   Date: _____   Section: _____

# Microscopic Observations

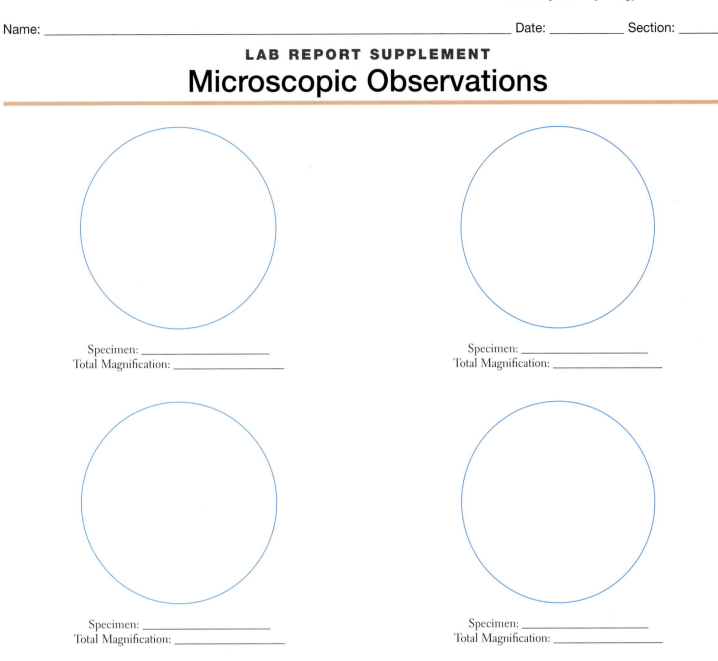

Specimen: _____
Total Magnification: _____

Specimen: _____
Total Magnification: _____

Specimen: _____
Total Magnification: _____

Specimen: _____
Total Magnification: _____

**Notes:**

Name: _____ Date: _____ Section: _____

# Microscopic Observations

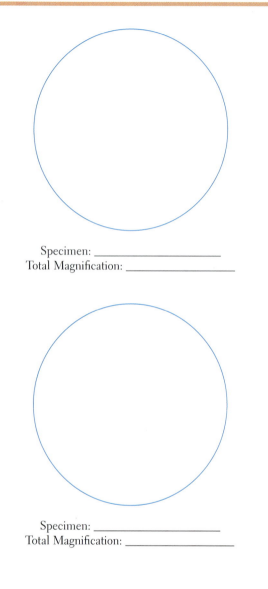

Specimen: _____
Total Magnification: _____

Specimen: _____
Total Magnification: _____

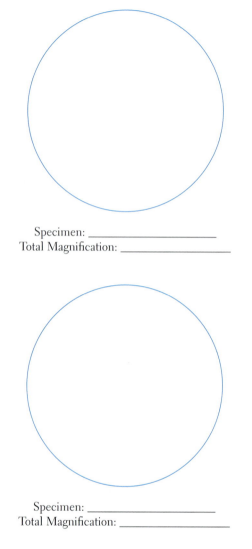

Specimen: _____
Total Magnification: _____

Specimen: _____
Total Magnification: _____

**Notes:**

Name: _____ Date: _____ Section: _____

# Microscopic Observations

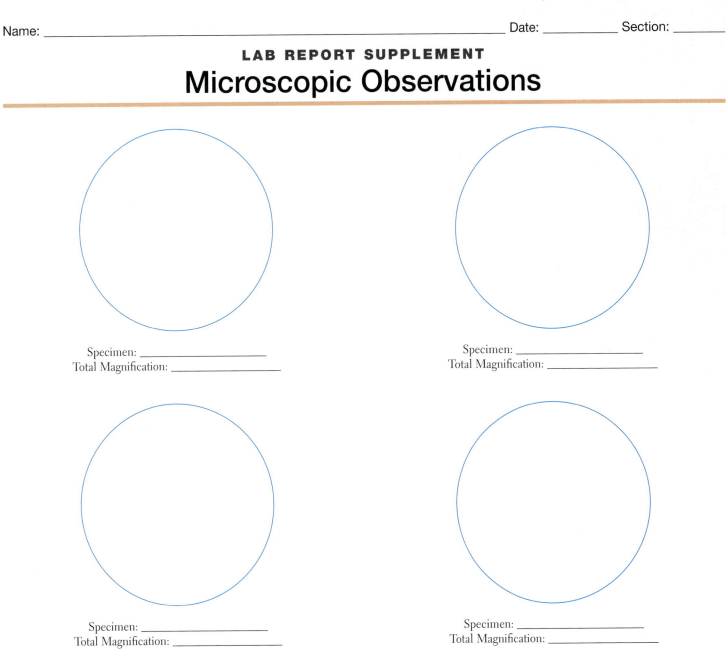

Specimen: _____
Total Magnification: _____

Specimen: _____
Total Magnification: _____

Specimen: _____
Total Magnification: _____

Specimen: _____
Total Magnification: _____

**Notes:**

Name: _____ Date: _____ Section: _____

# Microscopic Observations

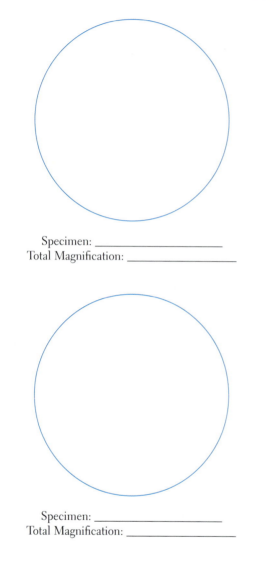

Specimen: _____
Total Magnification: _____

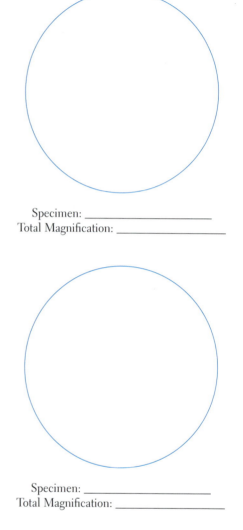

Specimen: _____
Total Magnification: _____

Specimen: _____
Total Magnification: _____

Specimen: _____
Total Magnification: _____

**Notes:**

Name: _____ Date: _____ Section: _____

**LAB REPORT SUPPLEMENT**
# Microscopic Observations

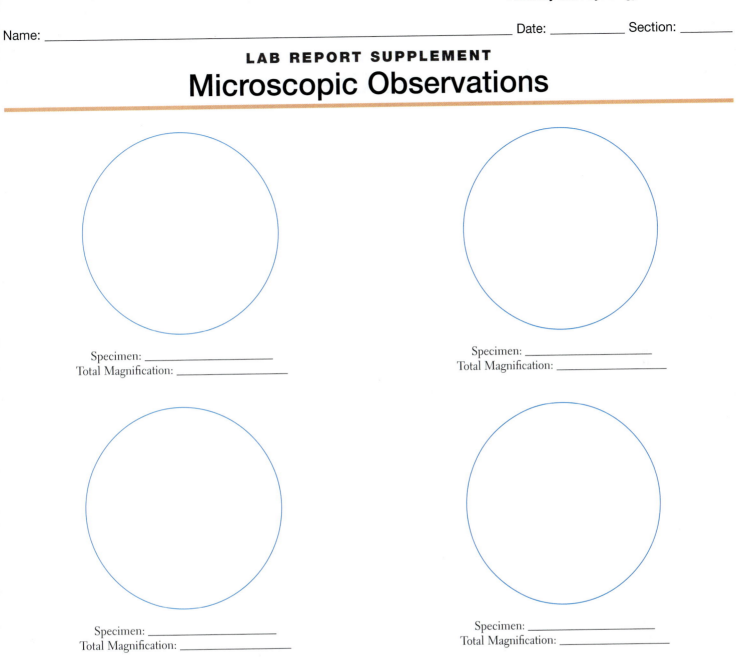

Specimen: _____
Total Magnification: _____

Specimen: _____
Total Magnification: _____

Specimen: _____
Total Magnification: _____

Specimen: _____
Total Magnification: _____

**Notes:**

Name: _____ Date: _____ Section: _____

## LAB REPORT SUPPLEMENT
# Microscopic Observations

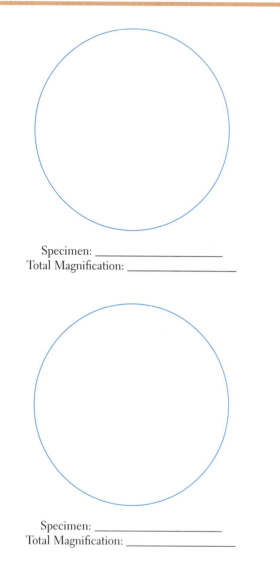

Specimen: _____
Total Magnification: _____

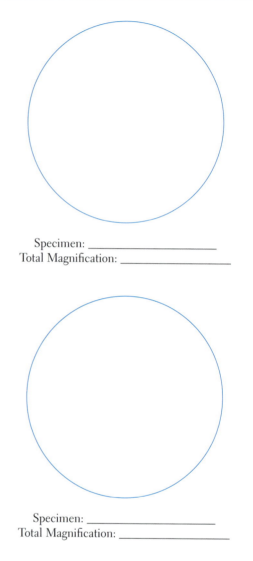

Specimen: _____
Total Magnification: _____

Specimen: _____
Total Magnification: _____

Specimen: _____
Total Magnification: _____

**Notes:**

Name: _____ Date: _____ Section: _____

# Microscopic Observations

Specimen: _____
Total Magnification: _____

Specimen: _____
Total Magnification: _____

Specimen: _____
Total Magnification: _____

Specimen: _____
Total Magnification: _____

**Notes:**

Name: _____ Date: _____ Section: _____

**LAB REPORT SUPPLEMENT**
# Microscopic Observations

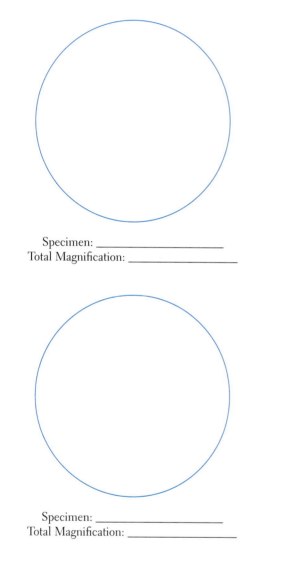

Specimen: _____
Total Magnification: _____

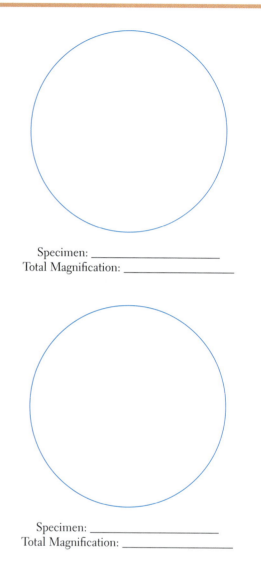

Specimen: _____
Total Magnification: _____

Specimen: _____
Total Magnification: _____

Specimen: _____
Total Magnification: _____

**Notes:**

Name: _____ Date: _____ Section: _____

**LAB REPORT SUPPLEMENT**
# Microscopic Observations

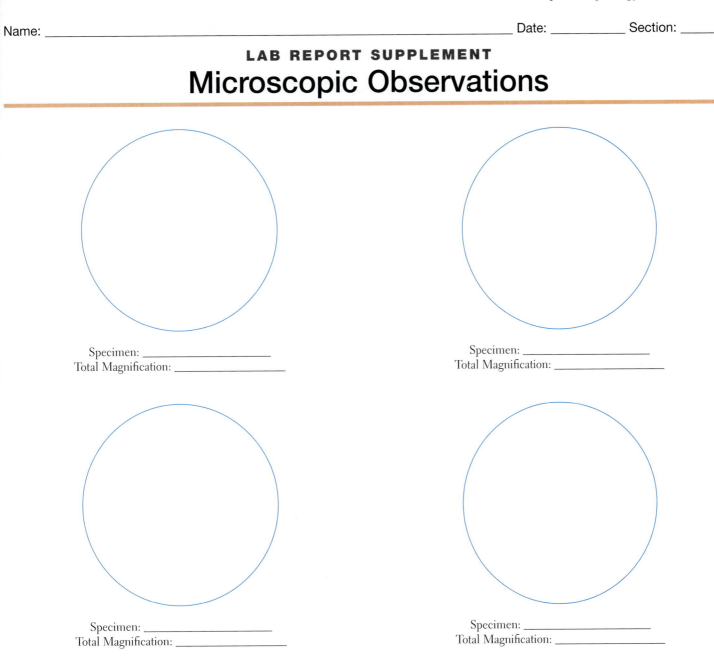

Specimen: _____
Total Magnification: _____

Specimen: _____
Total Magnification: _____

Specimen: _____
Total Magnification: _____

Specimen: _____
Total Magnification: _____

**Notes:**

Name: _____ Date: _____ Section: _____

# Microscopic Observations

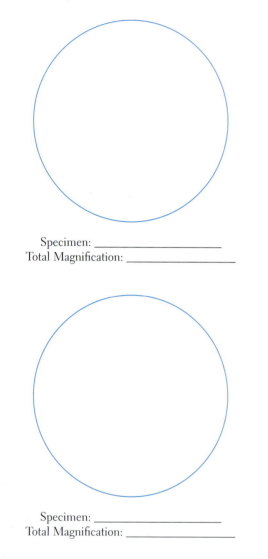

Specimen: _____
Total Magnification: _____

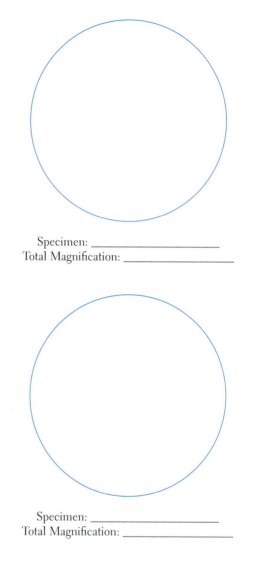

Specimen: _____
Total Magnification: _____

Specimen: _____
Total Magnification: _____

Specimen: _____
Total Magnification: _____

**Notes:**

Name: _____   Date: _____   Section: _____

# Microscopic Observations

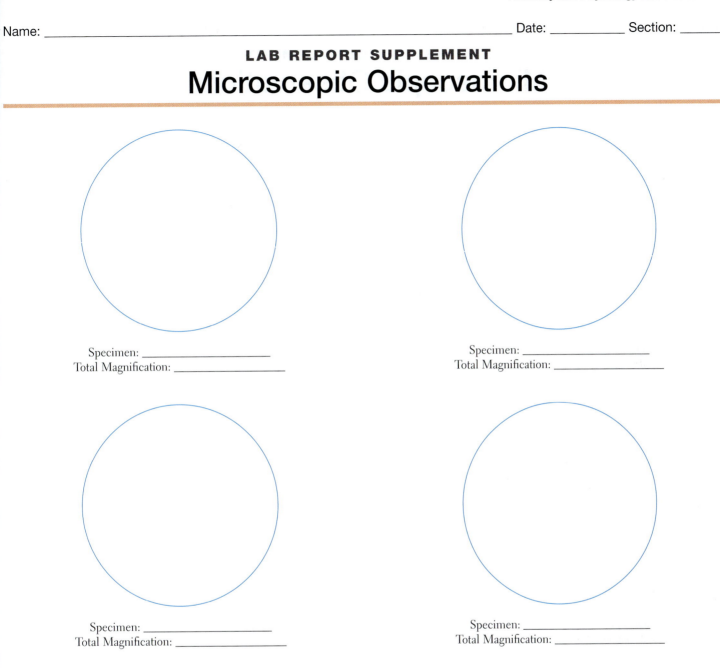

Specimen: _____
Total Magnification: _____

Specimen: _____
Total Magnification: _____

Specimen: _____
Total Magnification: _____

Specimen: _____
Total Magnification: _____

**Notes:**

Name: _____ Date: _____ Section: _____

**LAB REPORT SUPPLEMENT**
# Microscopic Observations

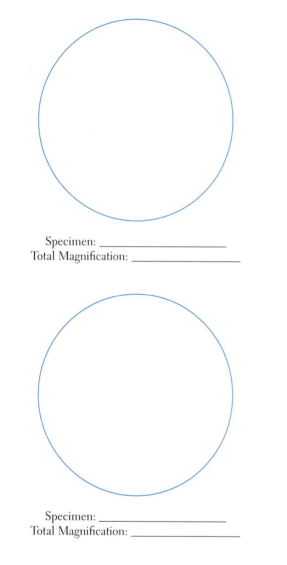

Specimen: _____
Total Magnification: _____

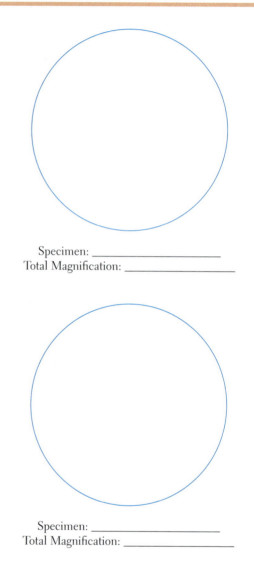

Specimen: _____
Total Magnification: _____

Specimen: _____
Total Magnification: _____

Specimen: _____
Total Magnification: _____

**Notes:**

Name: _____    Date: _____    Section: _____

**LAB REPORT SUPPLEMENT**
# Microscopic Observations

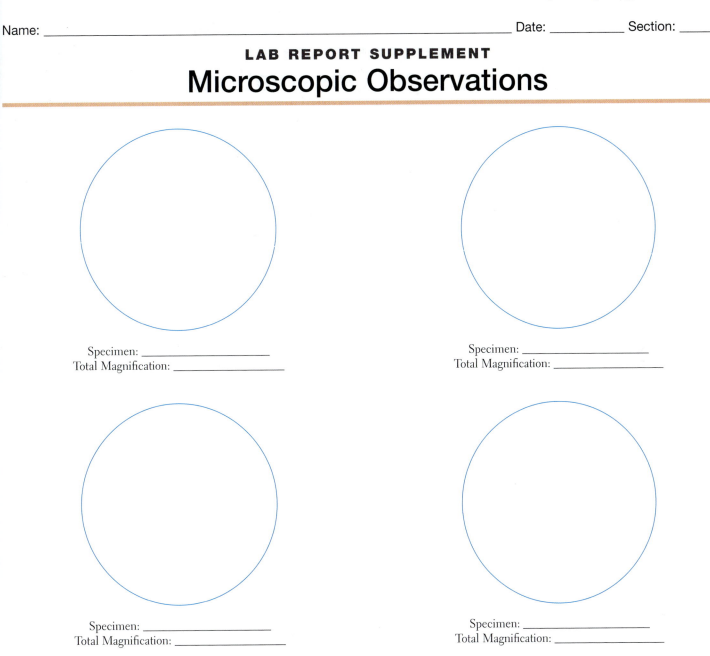

Specimen: _____
Total Magnification: _____

Specimen: _____
Total Magnification: _____

Specimen: _____
Total Magnification: _____

Specimen: _____
Total Magnification: _____

**Notes:**

Name: _____ Date: _____ Section: _____

# Microscopic Observations

Specimen: _____
Total Magnification: _____

Specimen: _____
Total Magnification: _____

Specimen: _____
Total Magnification: _____

Specimen: _____
Total Magnification: _____

**Notes:**